APPALACHIAN WILDFLOWERS

Appalachian Wildflowers

Thomas E. Hemmerly

THE UNIVERSITY OF GEORGIA PRESS

ATHENS AND LONDON

© 2000 by the University of Georgia Press
Athens, Georgia 30602
All rights reserved

Designed by Erin Kirk New
Set in Berkeley Oldstyle Medium by G&S Typesetters, Inc.
Printed and bound by Asia Pacific Offset

The paper in this book meets the guidelines for permanence
and durability of the Committee on Production Guidelines
for Book Longevity of the Council on Library Resources.

Printed in China

04   03   02   01   00   C   5   4   3   2   1

04   03   02   01   00   P   5   4   3   2   1

Library of Congress Cataloging-in-Publication Data

Hemmerly, Thomas E. (Thomas Ellsworth), 1932–
Appalachian wildflowers / Thomas E. Hemmerly.
p. cm.
Includes bibliographical references and index.
ISBN 0-8203-2164-8 (alk. paper).—ISBN 0-8203-2181-8
   (pbk. : alk. paper)
1. Wild flowers—Appalachian Region Identification.
   2. Wild flowers—Appalachian Region Pictorial works.
   I. Title.
QK122.3.H46 2000
582.13'0974—dc21   99-29638

British Library Cataloging-in-Publication Data available

To my wife, Beverly,

Who shared the joys (and trials)

of travel in the Appalachians

as in life.

Contents

Author's Notes

Amazon Basin, Black Forest, Caribbean Islands, Sea of Galilee, Point Pelee, Okefenokee Swamp, Mount Sinai, Rio Grande, County Limerick, and Appalachian Mountains—all these names and the special places they designate are among those that fascinated and intrigued me as a youngster. But the word *Appalachian* especially captured my fancy, even though those mountains were, for me, the most local of the world's wonders. Later, I learned that these mountains, though not nearly the highest, are among the oldest, the most beautiful, and the most visited in the world. I also became aware of their importance in the history of the New World; no other physical feature played a larger role in the history of North America than did the Appalachians. Certainly, these ancient and complex mountains with their diverse biota should be considered national and international treasures.

As a botanist and ecologist, I am concerned with the need to study and protect not only the biological diversity of the Appalachians but also that of other regions. Increasing human population pressures, industrialization, and the need for forest and mining resources put the world's flora and fauna at risk. Alabama native, now Harvard University biologist, Edward O. Wilson, in his book *The Diversity of Life*, said, "The folly our descendants will least forgive us is mass extinction of species." Vice President Al Gore (who also has Appalachian roots), in his landmark book *Earth in the Balance,* warns of the global ecological problems that must be solved in order for our grandchildren to inherit a livable and sustainable world. Fortunately, others, too, are sounding the call for the regional, national, and international concern and action necessary to enter the twenty-first century without major environmental catastrophes.

"Reading the landscape" was a recognizable skill long before modern conservationists began using the term "landscape ecology." Both mean the same thing: the ability to interpret nature not only in terms of discrete fragments, like forests, streams, and meadows, but also as a whole composed of dynamic units with indefinite boundaries interfacing, blending, and constantly affecting each other.

Unfortunately, modern education does not do a good job of helping people read the landscape. Even college biology majors are trained to think mainly in terms of blood and guts; cells, tissues, and organs; or molecules, atoms, and electrons. Too seldom are field studies required. Reed F. Noss, in a 1996 editorial in the journal *Conservation Biology,* laments the loss of field-oriented scientists. Even ecology, a science spawned by natural history, is being practiced increasingly by keyboard jockeys who "model" hypothetical ecosystems rather than experiencing them firsthand.

Of course, careers in the biological sciences require a firm grasp of animal architecture and a sound underpinning in organic chemistry and biochemistry; computers are indispensable for analyzing and storing data. But all persons, in order to be good citizens of planet Earth, should have a basic knowledge of their local natural areas and be able to relate them to the larger ecological picture, to read the landscape and respond appropriately to it.

As a result of the work of numerous scholars over almost three centuries, many manuals are available to help professional botanists read the landscape and, more particularly, identify Appalachian plants. For the less well trained, illustrated field guides are useful within certain mountain habitats (e.g., mountain timberlines), and local guides treat areas such as New England or particular states or provinces. But no single illustrated publication serves as a field guide to flowering plants throughout the Appalachians; thus, the justification for this volume.

This book is intended to be useful in helping you view wildflowers in the context of their environment, that is, to read their landscape. It attempts to go well beyond a simple identification of mountain plants; it should also be useful for learning about plant habitats, about factors that determine the distribution of plants, and about the many ways that plants are useful to humans. Such an appreciation should lead to environmental awareness, especially of the Appalachians and their ecology, but by extension to the entire global landscape, or biosphere, of which we are a part.

Acknowledgments

No author is more indebted to those who have gone before than myself. Long before I became a student of the flora of the Appalachians, the mountains had been explored, named, and mapped, and their long geological history outlined. Pioneer botanists with names such as Bartram, Mitchell, Clayton, and Gray had recognized and documented the uniqueness of Appalachian plants. These naturalists were followed by a new wave of scientifically trained plant taxonomists and ecologists of the twentieth century. Many of their works are listed in the bibliography.

My scientific interest in the plants of the Appalachians began in the late 1950s when my brother Ken and I camped in the Smoky Mountains while participating in several annual Gatlinburg Wildflower Pilgrimages. Later, serving as a field guide for these events, I shared the trails of the Great Smoky Mountains National Park with many notable field botanists. Deserving the most prominent recognition is the late A. J. "Jack" Sharp, long-time "dean" of Appalachian botanists and professor of botany at the University of Tennessee, Knoxville. Many others who shared their knowledge (all students or associates of Jack Sharp) include the late Frank Barclay and Herman O'Dell; also Ben Channell, Joe Chapman, Wayne Chester, Ed Clebsch, Hal DeSelm, William Ellis, Murray Evans, Dana Griffin, Alan Heilman, Fred Norris, Elsie Quarterman, Gwynn Ramsay, and Eugene Wofford.

At Middle Tennessee State University, financial assistance, which made travel and study possible, was provided by Summer Research Grants for 1992 and 1993. Colleagues who read and made suggestions for improving the manuscript or have otherwise made contributions include Kurt Blum (botany), Warner Cribb (geology), and Phil Mathis (genetics). Landon McKinney checked my identifications of violets. Andy Roadarmel, my grandson, provided field assistance on several mountain forays.

Ashley Messick deserves primary credit for her skillful preparation of the manuscript. Others involved include Lori Sain Smith, Sherry Messick, Emily Kee, and Marion Wells.

Charles Backus, director of the Vanderbilt University Press, and Bard Young, editor, were involved in the early stages of planning this book. The press graciously allowed the use of several photographs that appeared in my book *Wildflowers of the Central South.*

The staff of the University of Georgia Press deserves credit for the effective application of their respective skills that resulted in bringing this volume to completion. Included are: Karen Orchard, director; Barbara Ras, senior editor; Kristine Blakeslee, project editor; Kathi Morgan, production coordinator; Erin New, designer; and Anne Gibbons, freelance copy editor.

All photographs were made by me except Shale barrens, Greenbrier County of southeastern West Virginia, by Carol Baskin; Serpentine barren of Soldier's Delight, Maryland, contrasts with the Virginia pine forest, by Milo Pyne; *Trillium grandiflora,* by Jim Lea; and *Corallorhiza maculata,* courtesy Great Smoky Mountains National Park. Spencer Graves of Batey's Photography, Murfreesboro, Tennessee, provided useful photographic assistance.

The expense of printing the color plates was subsidized by a generous grant from Catherine Keever. A renowned plant ecologist, she is known, among her many other contributions, for pioneer studies into the mechanisms involved in old-field succession in the Piedmont region of the eastern United States.

Special recognition is due two eminent plant ecologists, both of whom have been a source of encouragement to me and support for the publication of *Appalachian Wildflowers.* Catherine Keever, professor emerita, Millersville State University, provided a generous grant toward the expense of printing the color plates. An authority on the ecology of forests of the eastern United States, she is especially known for her pioneer studies in old-field succession in the Piedmont region. My doctoral studies in plant ecology at Vanderbilt University were under the direction of Elsie Quarterman, now professor emerita. She and her numerous graduate students have contributed greatly to our appreciation and understanding of cedar glades, a unique type of ecosystem found primarily in middle Tennessee.

Using This Book

Wildflowers, like all living things, are constantly influenced by an ever-changing set of complex and interrelated factors, that is, their environment. Not only is the current environment important but also past ones extending back through centuries and millennia; those ancestors of life now on Earth were affected and selected by conditions in existence during those times. Thus in our attempt to better understand the plant life of the Appalachians, we are concerned with the formation of the mountains, the forces that have altered them, and the mantle of soil that covers their rocks—in short, everything pertinent to interpreting the mountain flora.

This book, therefore, begins with a consideration of humankind's relatively brief sojourn in the Appalachians and with the physical and ecological background of the region (chaps. 1 and 2). It then moves to an overview of the principal plant communities of the mountains (chaps. 3–5). This introduction to Appalachian environments serves as background for the second part.

The wildflowers illustrated in this volume are organized primarily by color and secondarily by the grand group to which they belong: monocots or dicots. Further explanation of this system is given in "Using the Color Plates."

To feature photographs of all the wildflower species (an estimated 3,000 to 3,500) of the Appalachian region would have resulted in a book much too large and costly to be practical. Obviously, a great deal of selectivity was necessary. In general, an attempt was made to include as many plants representative of Appalachian diversity as possible. Wildflowers that are showy and relatively common are more likely to be featured than inconspicuous uncommon ones. Because of their special interest, however, many infrequent to rare plants are also included. Several books (some of several volumes each) that provide a more thorough coverage of Appalachian wildflowers are listed in the bibliography.

The information opposite the photograph of each species of wildflower follows a reasonably consistent pattern. The common name (including the French name in parentheses for those plants found in French-speaking areas of the mountains) and scientific name are followed by the family name (the suffix "aceae" denotes a plant family). Nomenclature, including both family and scientific names, conforms closely to that of Gleason and Cronquist's *Manual of Vascular Plants* (2d ed.). For plants outside its region, *Gray's Manual of Botany* (8th ed.) by M. L. Fernald or Eugene Wofford's *Guide to the Vascular Plants of the Blue Ridge* were consulted. If a species is also widely known by a synonym or alternative scientific name (unfortunately, names are sometimes changed), that name is shown in brackets.

The narrative contains a brief description of the plant, its estimated relative abundance, habitat, and geographical distribution. The usual flowering time is also given. Keep in mind that flowering times may vary from place to place and year to year. In general, a given species will begin to flower first at lower elevations in the southern part of its range. In some seasons, favorable weather conditions may allow flowering to begin earlier or continue longer than indicated.

When applicable, information is provided on ethnobotanical, economic, or medicinal uses. One or more similar or related species may also be described briefly. Less common terms used are in the glossary (appendix 1).

To identify an unknown plant, you may wish to take one of these approaches:

1. If you think you know the name, scientific or common, use the index to locate the illustration to confirm the identification. It may be the one you thought or a related one.
2. If you don't have a clue, consult the appropriate color group in the table of contents and check that section of illustrations. Note the following explanations concerning the color sections: White includes cream and other pale tints. Some yellowish green flowers are in the green/brown section. Red includes rose and maroon; pink includes coral and mauve; blue/purple includes lavender, lilac, magenta, and violet; and green/brown includes other dark or indistinct colors.
3. If the plant is found in a special habitat (e.g., spruce-fir forest, cove hardwood forest, bog, or bald), see the list of plants for that habitat (chaps. 3–5).

Many opportunities exist throughout the Appalachians for "botanizing." Some of the best areas to visit are described in appendix 2.

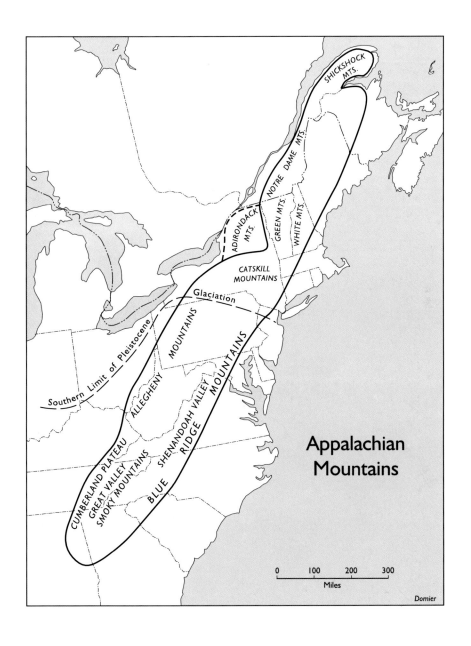

SHICKSHOCK MTS.

NOTRE DAME MTS.

ADIRONDACK MTS.

GREEN MTS.

WHITE MTS.

CATSKILL MOUNTAINS

Glaciation

MOUNTAINS

Southern Limit of Pleistocene

ALLEGHENY

CUMBERLAND PLATEAU

GREAT VALLEY

SMOKY MOUNTAINS

SHENANDOAH VALLEY

BLUE RIDGE MOUNTAINS

Appalachian
Mountains

0   100   200   300
Miles

Domier

Part 1

Ecology of the Appalachians

CHAPTER 1

People and Their Mountains

Geographers generally designate the Appalachians as the mountain chain that extends some 1,600 miles from the Gaspé Peninsula in Quebec to northern Alabama and Georgia. Because the Appalachians were formed as a part of the same grand series of geological events, they are virtually continuous. A generally favorable climate and an abundance of moisture throughout has resulted in a luxuriant growth of plants, which in turn sustain numerous and diverse animal populations.

Native Americans and Europeans

Following millennia without human life, Appalachia has undergone a relatively recent succession of occupations by a variety of ethnic groups. For several centuries, it was home to various Native American peoples from the Catawbas and Cherokees in the south to the Senecas, Penobscots, and Micmacs in the north.

Native Americans' detailed knowledge of plants—their identification, distribution, uses, even what we today would call their ecology—is now recognized as being of a much higher order than was thought by early European settlers. There is reason to believe that at the time of first European contact, the level of medicine as practiced within many of the Indian tribes was probably more advanced than that of European medicine. The abundance and diversity of Appalachian plants, as well as the ingenuity of the practitioners, were no doubt responsible for their successes. Indian medicines and accompanying rituals were not always efficacious, but they were less likely to be as dangerous as blood-letting or the use of poisonous mercury and arsenic compounds, which were common practices in Western medicine.

Eventually, Europeans settled in the Appalachians and brought with them a general knowledge and appreciation of plants as well as some Old World plants, which they cultivated and used. Despite the uniqueness of many mountain plants, the settlers related them to those of the Old Country with which they were familiar. Thus they had some guidelines to indicate possible uses of plants for food, medicine, dyeing, and other purposes. In this pre-industrial economy, wood from many species of trees was a plentiful raw material from which were fashioned an almost limitless supply of items, including furniture, tools, boats, fences, and buildings.

The response of Western physicians and scientists to American Indian medicine was mixed. Initially, and for an extended period, they were skeptical, often relegating Indian herbs and practices to folklore and superstition. Increasingly though, throughout the nineteenth and early twentieth centuries, various ethnobotanists and historians came to recognize, document, and appreciate the contributions of Native Americans. The eminent historian of Indian medicine Virgil J. Vogel counted 220 herbal substances listed in the *U.S. Pharmacopeia and National Formulary* between 1800 and 1965 that had been "borrowed" from Indians.

The Bartrams and Company

A much smaller but important group of Europeans had a different reason for wanting to learn about Appalachian plants. These were the early explorer-naturalists, who were overwhelmed by the great number and diversity of plants in the mountains.

In a letter to Carolus Linnaeus, Alexander Garden (for whom the shrub gardenia is named) wrote on 22 March 1756, "The hopes of returning richly laden with the spoils of Nature of our Apilachian [sic] Mountains is more to me than soft ease or hopes of sordid gain."

But it was John Bartram (1699–1777) and his son William (1739–1823) who were the most important and best known of the eighteenth-century naturalists of the American colonies. The father had operated a thriving nursery business in Philadelphia; his plants and seeds shipped to England and other European countries introduced many New World plants to Europe.

The Bartrams traveled widely throughout the colonies, often venturing into the Appalachians. They made voluminous notes on fossils, minerals, and animals, but observing and collecting plants was their primary goal. As John Bartram said, "O Botany! delightfulest of all sciences, there is no end to its gratifications."

Continued international interest in Appalachian plants was stimulated by several editions of William Bartram's *Travels,* which was translated into French, German, and Dutch. Subsequently, many explorer-botanists traveled to the mountains. Among them was France's André Michaux, who collected plants for the gardens of Versailles. The Scotsman John Fraser, often a traveling companion of Michaux, is remembered by Fraser fir (*Abies fraseri*).

By the early part of the twentieth century, improved roads made the mountains more accessible to a greater number and variety of people, including developers, entrepreneurs, and tourists. Since then, neither the mountains nor the mountain people have been the same.

Plate Tectonics

Pointing to a world map, "Miss Douglas," my fourth-grade teacher, called the attention of the class to the shapes of the African and South American continents. Even though separated by the Atlantic Ocean, they appear to fit together like pieces of a jigsaw puzzle. The theory of continental drift postulates that continents have been, and indeed still are, moving. This would explain the complementary shapes of various continents now thousands of miles apart.

This earlier, vague notion of continental drift is embodied in the modern theory now known as plate tectonics; this revolutionary principle serves as the basis for modern geological interpretation. It furnishes a mechanism, as well as evidence, for the earlier speculation, which was based merely on circumstantial evidence, that continents are parts of plates that drift slowly on a mantle, solid but able to flow because it is near its melting point. So for the first time, we have answers to the question of Earth's restlessness: explanations for what causes earthquakes and volcanoes, and how mountains are formed.

The Appalachian Revolution

Geologists began serious studies in North America during the nineteenth century. One of their first (and toughest) challenges was the Appalachians. The consensus was that the mountains had been formed by contraction of the earth's crust as it cooled, rather like an apple skin that becomes wrinkled as the apple ages and dries.

Today, plate tectonics provides a much more plausible explanation for mountain formation. The Appalachian scenario goes like this: About 570 million years ago, a large land mass of the Northern Hemisphere broke in two, forming the early continents destined to become North America and Africa; between them was the primeval Atlantic Ocean. Next, the two continents began to converge, a process lasting some 300 million years and resulting in a "collision" between them. From this persistent contact was produced a buckling along the eastern edge of North America, which was the young Appalachians.

This revolution, which occurred over a period of some 250 million years, consisted of several separate episodes, or orogenies. Later, compressions and continued contact of the continents forced the mountains more than 150 miles westward. The topsy-turvy arrangement of rock strata within the mountains today reflects this chaotic geological history. Other mountain ranges, such as the European Alps and those of western North America, were formed more recently than the Appalachians, apparently by comparable geological processes. The Appalachians have been subjected to relentless forces of change that have drastically reduced their height and changed their contours.

The Shaping of the Mountains

Mountains are made to be worn down; the Appalachians are no exception. As they now exist, these mountains are the product of some 200 million years of continuous erosion combined with intermittent glaciation.

To the casual observer, the rock that underlies the mountains is the most stable feature of the landscape. These rocks, though, are vulnerable to the constant forces of erosion due to weathering processes. As masses of cold, dry air from the north clash with humid, warmer air from the south, the mountains are subjected to heavy rains and snows. Frequent hard freezes, alternating with thaws, also have an effect, especially in the northern mountains. The ice within rock crevices acts to repeatedly divide large rocks into smaller ones.

By such processes, the mountains have been greatly reduced in elevation, and their former jagged shapes have been rounded. By contrasting the

Seneca Rocks, in eastern West Virginia, stand as an ancient sculpture reflecting the tumultuous geological past of Appalachia.

Rockies with the Appalachians, we can appreciate the transformation from young to ancient. Another result of the weathering of rocks is the formation of soil, which is necessary for the establishment of plants.

But other profound changes were in store for the Appalachians. During a more recent geological epoch—the Pleistocene, or Ice Age, which began a little less than 2 million years ago—the northern mountains were covered with ice. Although some present arctic glaciers are quite large, they are puny indeed in comparison to the mile-thick Pleistocene ice sheet that blanketed the northern portion of the continent. Although all the Appalachians were subjected to cooling, the glaciers extended only as far south as New York or northern Pennsylvania. Actually, there were several glacial cycles, each including warmer interglacial periods. The last glacial episode, the Wisconsin, ended only about 10,000 years ago.

Geology and Biogeography

What is the importance of this tumultuous geological past in our understanding of the flora of the Appalachians? Glaciation and erosion, together with other forces acting on the mountains, were responsible for the varied landscape, including many potential microhabitats that, in turn, were necessary for the development of the grand botanical mosaic of the Appalachians.

Because the southern Appalachians have been neither submerged under the sea nor covered by glaciers, they have gained a reputation as the "cradle of vegetation" for eastern North America. It is now recognized that, although not glaciated, the area was subjected to greatly reduced temperatures, resulting in the migration of its flora several hundreds of miles southward. According to the "river corridor" hypothesis of University of Tennessee professors P. A. Delcourt and H. R. Delcourt, plants moved southward, perhaps two hundred or more miles, along river valleys during glacial epochs, returning northward along these routes as the Pleistocene climate warmed. Thus it appears that areas containing cove hardwood forests, noted for their complexity and diversity, have not, as was long believed, served continuously as havens for the numerous species now found there. The richness of these communities is better explained by the combination of copious rainfall, fertile soils, and relatively long growing seasons.

The Appalachians Today

Botanists often distinguish between the northern and southern Appalachians. Some would allow the line of demarcation to coincide with the extent of Pleistocene glaciation, whereas others use the east-flowing Roanoke River of southern Virginia. This book recognizes three mountain regions: the southern Appalachians, the portion south of the Roanoke River; the northern

Mount Albert, one of the highest peaks of Quebec's Shickshock Mountains, has alpine and serpentine communities, as well as old-growth deciduous and boreal forests.

Appalachians, the glaciated portion; and the middle Appalachians, which link the other two.

In Quebec, the northernmost portion of the Appalachian system, the Shickshock Mountains rise nearly 4,000 feet above the eastern banks of the Saint Lawrence River. This system, composed of sometimes separated and sometimes overlapping ranges, extends some 2,000 miles to the south. Even higher elevations (above 6,000 feet) are characteristic of the southern Appalachians, especially near the borders between Tennessee, North and South Carolina, and Georgia. But the southern terminus is less dramatic than the northern: southwest of the Smokies of Tennessee and North Carolina, the mountains gradually diminish in elevation and fade into the Piedmont and Coastal Plain of Georgia and Alabama.

In the northern Appalachians, the Notre Dame Mountains are just to the south of the Canadian Shickshocks. In the United States the White Mountains practically cover New Hampshire and include Mt. Washington at 6,288 feet, the highest peak in the Northeast. The Green Mountains of Vermont extend into western Massachusetts, where they are called the Berkshires, and into western Connecticut as the Litchfield Hills.

To the east, in upper New York, are the famous Adirondacks, which include extensive old-growth forests and an essentially Appalachian flora. For this reason they are included within our area, despite their geological ties to the Canadian shield of southcentral Canada and the Great Lakes region. In contrast, the Catskill Mountains of southeastern New York are geologically, as well as botanically, an integral part of the Appalachians.

The Allegheny and Blue Ridge Mountains have their beginnings in Pennsylvania; farther south there is a divergence of these mountains in northern Virginia, with the Shenandoah Valley separating the Alleghenies to the west from the Blue Ridge to the east.

The Blue Ridge Mountains in North Carolina and Tennessee include, from north to south, the Black Mountains of northwestern North Carolina and the Smoky Mountains, which straddle the Tennessee–North Carolina border. Among the notable peaks of the southern Blue Ridge are Grandfather Mountain (5,938 ft.); Roan Mountain (6,267 ft.); Mt. Mitchell (6,684 ft.), the highest peak of the Appalachians; and Clingman's Dome (6,643 ft.), the highest peak of the Smokies.

The Alleghenies continue southward from Pennsylvania, becoming the Cumberland Plateau, which extends through eastern Kentucky and Tennessee. The Shenandoah Valley continues from Virginia into Tennessee and northern Alabama, where it is called the Great Valley. The southernmost part of the valley separates the southern Blue Ridge, which extends into northern Georgia, from the Cumberland Plateau, which ends in northern Alabama.

CHAPTER 2

Dynamics of Mountain Ecosystems

If all nature could be characterized by a single word, that word would, undoubtedly, be *change*. Geological changes such as those wrought in the region now called the Appalachians (see chap. 1) set the stage for biological changes via evolution, the process by which species undergo not only incremental internal changes but also the more drastic and profound process of speciation, by which new species arise from preexisting ones. In this way biodiversity increases, and so does the complexity of the biotic communities and ecosystems that embrace these species.

Ecosystem Development

An ecosystem is a local unit of nature. Whether a pond, a forest, or a mountain bald, each consists of all its living inhabitants and their physical environment. Ecosystems are dynamic units that change slowly at times and more rapidly at others. One of the major objectives of community ecologists —especially those of North America during the last hundred years—has been to document these changes, to develop a terminology to describe them, and to theorize as to the underlying causative mechanisms. In 1898 Henry Cowles pioneered the theory of ecological succession: ecosystems (he called them communities) pass progressively through predictable stages before becoming a permanent, or climax, ecosystem. Such a climax is determined primarily by the prevailing climate of that region.

During the 1920s some plant ecologists began to develop other ways of interpreting the dynamics of ecosystems. Notable among them was Henry Gleason, who attacked the earlier view and replaced it with his "individualistic concept": ecosystems change with time but in a more unpredictable, less deterministic manner, and molded by a combination of factors rather than by climate alone. More recently, Frank E. Egler, Catherine Keever, Frank McCormick, Rebecca Sharitz, and others have conducted field experiments to identify the mechanisms that drive ecosystem development (as succession is now called).

Like all ecosystems, this Adirondacks (New York) beaver pond includes a diversity of plants, animals, and microorganisms.

Following a forest fire in Maine's Acadia National Park, succession has produced an almost pure stand of paper birch (*Betula papyrifera*).

Actually, ecosystems are probably so complex that no single paradigm is sufficient to explain the dynamics of all types of ecosystems. But in order to fully appreciate a mountain forest, field, or bog, one should know its past history and use as a predictor of future changes the established pattern of ecosystem development for that region.

Soils Are More than Dirt

According to Hans Jenny, eminent pioneer of modern soil studies, "It is embarrassing not to be able to agree on what soil is." Although we might not find it easy to define soils, we recognize them as complex life-sustaining systems. They vary greatly from region to region as a result of the conditions, especially the climate but also the vegetation, under which they have developed.

Mature soils, found throughout the mountains, are composed of horizons (layers). These are visible in a soil profile, a vertical slice such as can be seen in a road cut. At the surface of the profile is the O-layer (organic) of undecomposed matter such as recently fallen leaves. Just below is the A-layer (topsoil) and, under it, the B-layer (subsoil). Below the B-layer is bedrock, the nature of which also affects the soil-forming process.

In the Appalachians are found three principal mature soil types, each supporting and resulting from a major type of forest (see chap. 3). Spodosols support evergreen boreal forests. These acidic soils have a thick O-layer, composed of decomposing needles, a light gray A-layer and a darker B-layer. Alfisols are associated with deciduous forests. Having less organic matter, they are less acidic. Both A- and B-layers are dark (dark gray to brown) and with less contrast in the appearance of the two layers. Ultisols develop in certain low-elevation forests of the southern Appalachians where the climate is warmer and more humid than it is farther north. They are more intensely weathered, leaving them red or yellow. They are also less fertile and have a higher clay content than alfisols.

Less mature, azonal (lacking stratification) entisols are also found in the mountains. They are typically associated with disturbed areas (see chap. 4) or wetlands (see chap. 5).

Climates and Microclimates

Most of the eastern United States, including nearly all the Appalachians, lies within the climatic region described as humid mesothermal. As this term indicates, our mountains have a moderate climate, both in terms of rainfall and temperature.

In these mountains, considerable microclimatic differences—that is, local departures from the general climate—exist between north-facing slopes and south-facing slopes. South- and southwest-facing slopes receive the most solar energy, causing them to be both warmer and drier; this is especially true in early spring before the leaf canopy reduces the amount of sunlight that reaches the forest floor. Also to be considered in the middle and southern Appalachians is the fact that north-facing slopes are, on average, steeper and include more minor irregularities (e.g., crevices) than south-facing ones, which are smoother and less steep.

These and other microclimates, such as those of depressions, wetlands, rock outcrops, and other specialized habitats, must be considered in order to understand the actual conditions under which plants exist.

The Mountain Calendar

Another feature of the Appalachian area is its well-marked seasonality: the year is divided into four distinct seasons, each characterized by constantly changing, yet somewhat predictable, events.

Spring in Appalachia arrives first at the lowest elevations of the southern-most mountains. Warmed by sun penetrating the naked branches of the trees, a riot of flowering herbs erupts on the forest floor, beginning typically in early March. From there the season progresses upward and northward, slowly at times but more rapidly during extended warm periods. Spring does not reach parts of New England until late April.

Even before spring events have been concluded in the north, the summer season is well underway in the south. There, on the forest floor, now shaded by a canopy of leaves, flowering continues, but at a reduced intensity. Many summer-flowering plants are trees or shrubs, or herbs in meadows, along roadsides, or in other more open habitats. In summer, rainfall is generally reduced; droughts may occur locally but are seldom severe or prolonged.

Autumn in the Appalachians is characterized by the brilliance of its deciduous tree and shrub foliage. Shorter days combined with cool night temperatures trigger hormonal signals within the leaves of deciduous plants, causing the formation of an abscission (separation) layer at the base of each petiole (leaf stalk). Chlorophyll concentrations are gradually reduced as less moisture is provided to each leaf. Soon the leaves will be shed, but before then subtle chemical changes within them cause various pigments to become enhanced. Among the pigments responsible for autumn leaves are carotenes (orange, red), xanthophylls (yellow), anthocyanins (red, purple), and tannins (brown).

Winters are relatively brief and mild in the southern Appalachians, especially at lower elevations; they are longer and harsher in the north and at

13

Autumn foliage,
at Fall Creek Falls,
Fall Creek Falls
State Park,
Tennessee.

higher elevations. Winter is not only a time of lessened activity for plants but also an opportunity for restoration and preparation for the advent of spring.

Thus all living things are affected by various interrelated environmental factors. As Rose Houk says in her delightful book *Great Smoky Mountains*, "The brook trout that swim in the pure cold streams depend on the shade of the rhododendrons, which in turn thrive on the acidic soils created partly by the bedrock."

CHAPTER 3

Mountain Forests

Interpreting Forests

Forest plants are arranged into horizontal layers. The tallest trees constitute the canopy; those not able to reach that layer form an understory (or sub-canopy) layer just beneath. Below this is the shrub layer, and on the forest floor are mosses, ferns, and wildflowers, which comprise the herbaceous layer.

In the Appalachians, there are two main forest types: deciduous and evergreen. The former, found in valleys and at all but the highest southern peaks, extend northward at increasingly lower elevations. Evergreen forests cover (or originally covered) most of the remaining mountain regions. In the south, they are found only at the highest peaks; northward, they are seen at ever-decreasing elevations. In northern New England and Quebec, evergreen forests that clothe the lower mountain slopes are continuous with those of adjacent lowlands. Evergreen forests are dominated by trees such as pine, spruce, fir, and hemlock.

At some places in the Appalachians, these two types of forests interface yet remain relatively distinct. More often, however, there is a gradual merging or mixing of the two to form transitional ecotone forests, which combine many of the species found in each of the adjacent types.

Reflecting climatic and other differences within the deciduous forests of the Appalachians, several forest associations may be recognized. Following the recent designations in *Biodiversity of the Southeastern United States: Upland Terrestrial Communities,* edited by William H. Martin and others, and based on the tree species of the canopy, these associations are mixed mesophytic, Appalachian oak, oak-hickory-pine, and northern hardwoods.

Mixed Mesophytic Forests

This forest region is found primarily on the Cumberland Plateau, from northern Alabama through Tennessee and Kentucky, with extensions into the Allegheny Mountains of western West Virginia and adjacent Ohio; outliers

Mixed forest (ecotone), middle elevations of Great Smoky Mountains National Park, Tennessee; trees of the boreal forest mix with those of the deciduous forest.

(noncontinuous portions) of the mixed mesophytic region occur as cove hardwood forests.

Its canopy is dominated by more than a dozen tree species including American Beech, Tulip-tree, Cucumber Tree, White Basswood, Sugar Maple, Yellow Buckeye, Red Oak, White Oak, and Eastern Hemlock. The understory includes Fraser's-magnolia, Allegheny Serviceberry, Redbud, Flowering Dogwood, Alternate-leaved Dogwood, American Holly, Mountain Holly, and Striped Maple. The shrub layer includes Pepperbush, Mountain Laurel, Dog-hobble, Strawberry Bush, and many others. Among the woody vines are Dutchman's-pipe, Virginia Creeper, several species of grape, and Poison Ivy.

Although the canopy tree species define a forest region, the herbaceous layer of the mixed mesophytic forest is of special interest because of its species richness. The flowering plants are especially showy in late spring before leaves of the woody plants overhead unfold. A few of these herbs are Doll's-eyes, Hepatica, May-apple, Dutchman's-breeches, Spring Beauty, several violets, Foam Flower, Bishop's-cap, toothworts, Mountain Phlox, Fringed Phacelia, Partridge Berry, Jack-in-the-pulpit, Solomon's-plume, Fraser's-sedge, Trout-lily, and several trilliums.

In summer, common herbs in flower include White Wood Aster, Blue Cohosh, Canada Waterleaf, Pale Jewelweeds, Spotted Jewelweeds, Tall Bellflower, coneflowers, White Snakeroot, and Solomon's-seal.

Cove hardwood forests of the southern Appalachians are especially lush and possess an unusually high biodiversity. Although geographically disconnected from the mixed mesophytic forest region, they represent this type of forest par excellence. Cove hardwood forests are found in protected valleys (coves) at low to middle elevations (to 4,000 feet), where the soil is deep and fertile and rainfall is ample.

In addition to the tree species listed for the mixed mesophytic forests, those found in cove hardwood forests include Bigleaf Magnolia, Eastern Hemlock, Mountain Maple, Yellowwood, and Carolina Silverbell. More than a thousand herbaceous species are found in cove hardwood forests, and they put on a spectacular spring show.

Appalachian Oak Forests

The Appalachian oak forest region includes the Blue Ridge Mountains and Great Valley and extends northward into much of Pennsylvania and southern New England. Today this great forest is quite different than it was a century ago. Earlier this century, a fungal blight of the American Chestnut caused by *Endothia parasitica*, introduced originally into New York, swept the mountains. The result was the virtual annihilation of the species.

The loss of this giant from the mountains obviously altered the tree composition of its forests, but succession has filled the gaps left by dead chestnuts with other tree species, especially their relatives, the oaks.

Among the most important canopy species of the Appalachian oak forest is White Oak, a widespread and abundant tree of the Appalachians. Other important oaks are Northern Red, Black, and Chestnut. Hickories include Pignut and Mockernut. On moister, or more mesic, sites are found Sugar Maple, Tulip-tree, Shagbark Hickory, and Bitternut Hickory. Within the Appalachian oak forest region, local site factors such as topography and elevation often result in major forest types that depart from the general idealized type. Thus one may encounter on xeric (dry) sites, oak-pine or chestnut-oak forests; on mesic sites, mixed mesophytic forests; and scattered throughout, oak-hickory forests. The presence of pines, Eastern Red Cedar, Quaking Aspen, or Bigtooth Aspen is evidence of past disturbances.

Shrubs may include Sweet Shrub, Great Rhododendron, Mountain Laurel, Highbush-blueberry, and Wild Hydrangea.

Herbaceous species include Bird-foot Violet, Trailing Arbutus, Galax, Spotted Wintergreen, Solomon's-plume, Yellow Trillium, and Pink Moccasin Flower.

Oak-Hickory-Pine Forests

The oak-hickory-pine forest region occupies primarily the Piedmont to the east and south of the Appalachians; also included is a portion of the Appalachian Plateau from southern Pennsylvania for some 150 miles along the

An oak-hickory-pine forest in eastern Kentucky.

Virginia–West Virginia border. Stands of oaks mixed with hickories and pines are also regularly found on ridges and south-facing slopes of the southern Appalachians to an elevation of nearly 4,500 feet. Soils of this forest region are not as moist as those of mixed mesophytic forests.

Canopy species of oak-hickory-pine stands in the Appalachians include White, Chestnut, Black, and Scarlet Oaks. Hickories include Pignut, Mockernut, and Bitternut. Pines include Pitch, Table Mountain, White, and Virginia. Without fire, pines are eventually replaced by various oaks. Somewhat less abundant than the oaks are Red Maple, Sourwood, Allegheny Serviceberry, Black Locust, and Sassafras. Tulip-tree may be more important than has been previously recognized.

The understory is dominated by tall shrubs, especially Mountain Laurel, but also Pepperbush, Great Rhododendron, Flame Azalea, Highbush-blueberry, and Deerberry. In places, a second lower layer is formed by other ericaceous shrubs such as Huckleberry and Hillside-blueberry.

On the forest floor may be found May-apple, Trailing Arbutus, Galax, Wintergreen, Goat's-rue, Trout-lily, and Pink Moccasin Flower.

Northern Hardwood Forests

This transitional forest region is the northernmost (and uppermost) of the deciduous forest areas. As such, it combines elements of adjacent boreal forest (see chap. 4) with those of the deciduous forests. The northern hardwood forest region extends from west of the Great Lakes to New England and New York, and from there southward into the southern Appalachians, where it is found at elevations just below the spruce-fir forests. Each "island" of spruce-fir in the middle and southern Appalachians is surrounded by northern hardwoods. In New England, northern hardwood forests are also found below evergreen forests but occur there at lower elevations and on adjacent lowlands.

The canopy of the northern hardwoods is dominated by three principal species: Yellow Birch, American Beech, and Sugar Maple. Other hardwoods include Northern Red Oak and Paper Birch. The important conifers are Red Spruce, Black Spruce, Balsam Fir, Eastern Hemlock, and White Pine. Eastern Hemlock is found on cooler, more mesic sites, whereas White Pine occupies drier, more exposed ones. The latter, together with Fire Cherry and Gray Birch, are indicators of past fires or other disturbances.

Understory trees and shrubs include Mountain Laurel, American Mountain-ash, Striped Maple, and Mountain Maple.

On the forest floor, herbaceous species include Big Yellow Wood-sorrel, Wild Bergamot, Wood Lily, Painted Trillium, and Pink Moccasin Flower. Many of the herbaceous species found here are also characteristic of other deciduous forest types and also of the boreal forest.

The northern hardwoods of the southern Appalachians extend on rich, well-drained soils to higher elevations than those further north. Here, Yellow Buckeye and White Basswood can be added to the canopy list. Shrubs include those listed above plus Wild Hydrangea, Alternate-leaved Dogwood, and Witch-hobble. Many of the spring wildflowers, especially where beeches are abundant, are those of cove hardwood forests.

Northern and Middle Spruce-Fir Forests

Traveling north within the lowland northern hardwood forests of northern New England and southeastern Canada, the percentage of deciduous trees is gradually reduced as they are replaced by ever-increasing numbers of evergreen conifers, cone-bearing trees with needle-like leaves. After passing through this ecotone of mixed forests, one is now in a very different kind of forest. Some call it the northern evergreen forest; others, the boreal forest. To recognize the dominant forest trees, we shall call them spruce-fir forests.

Spruce-fir forests extend across glaciated North America in a broad band, forming an arc from Newfoundland to Alaska. Included are the northern Great Lakes region and the northern portions of New York and New England. From Quebec and New England, they continue southward as isolated "islands" atop only the highest elevations of Virginia, West Virginia, North Carolina, and Tennessee.

In these spruce-fir mountain regions, growing seasons are short; winters, long and cold. Snows are frequent, blanketing the trees and forest floor for extended periods. The conical shape of the trees allows them to withstand the heavy loads of snow better than trees of other shapes.

Soils of the region, which are products of decomposing evergreen needles and affected by recent glaciation as well as the prevailing climate, are highly acidic, gray spodosols (see chap. 2). Compared to the mature, rich soils of nonglaciated deciduous forests, those of boreal forests are young and nutrient-poor.

The two most important dominant tree species of northern and middle old-growth or climax spruce-fir forests are White Spruce and Balsam Fir. Evergreen trees of secondary forests include Black Spruce and Eastern Hemlock, together with White, Red, and Jack Pine. Also present are these deciduous trees: Paper Birch, Quaking Aspen, Bigtooth Aspen, and Balsam Poplar. Pure stands of Paper Birch are often seen, especially in northern New England and New Brunswick after a disturbance.

Because of the dense, year-round shading by the canopy, an understory is seldom present. Also, especially as compared to deciduous forests, there is a reduction in the shrub layer; shrubs are successional or are seen along forest edges, where more light is available. Shrubs may include Green Alder, Bearberry, Sheep Laurel, Mountain Holly, and Common Elderberry.

Herbaceous species, too, are much less abundant than in deciduous forests. Look for wildflowers in forest openings, edges, or successional stands. Among spring wildflowers are Bunchberry, Twinflower, Canada Mayflower, Clinton's-lily, and Nodding Trillium. Along roadsides and disturbed sites in summer are Fireweed and Orange Hawkweed. In fall, there are several asters, especially the Large-leaved Aster.

Southern Spruce-Fir Forests

The southernmost extension of the boreal forest is in the form of isolated spruce-fir stands on high mountain peaks of the southern Appalachians, especially along the border between North Carolina and Tennessee. Those of the Smokies are perhaps the best known. Compared to their northern counterparts, these southern forests receive more precipitation (much in the form of fog) and have somewhat less severe winters; thus trees and other plants, such as mosses, grow more luxuriantly. Not surprisingly, millennia of separation from northern spruce-fir forests have produced both floristic and vegetational differences.

But the essential common characteristics of these forests become evident when hiking through them. The pungent aroma of the needles and the spongy feel of the organic mat blanketing the ground are the same. In these forests one notices not only the darkness of the interior due to the thick branches of the canopy trees but also the openness between the canopy and the ground. Wildflowers are sparse on the heavily shaded forest floor.

The two canopy dominants of the southern forests are Red Spruce and Fraser Fir. Firs have erect cones and white stripes on the underside of the needles; spruces have pendant cones and lack the stripes on their needles. These two species replace White Spruce and Balsam Fir of the boreal forests farther north. (Red Spruce is also found in the northern and middle

A branch of red spruce (*Picea rubens*) bears pendant cones.

Fraser fir (*Abies fraseri*), endemic to high elevations of the southern Appalachians, has erect cones.

Balsam fir (*Abies balsamea*) is seen here on Cadillac Mountain, Acadia National Park, Maine.

forests but not as a forest dominant as it is in the south). Other canopy trees are Allegheny Serviceberry, Yellow Birch, Fire Cherry, and American Mountain-ash.

Understory trees include Mountain Holly, Striped Maple, and Mountain Maple. Among the shrubs are Mountain Rosebay, Piedmont Rhododendron, Highbush Blueberry, Witch-hobble, Bush-honeysuckle, Red Elderberry, Alternate-leaved Dogwood, Southern Wild-raisin, and blackberries.

The herbaceous layer is richer than its northern and middle Appalachian counterparts. Spring-flowering plants include Spring-beauty, Common Wood-sorrel, Cuthbert's-turtlehead, Bluet, Jack-in-the-pulpit, Red Trillium, Painted Trillium, Indian Cucumber, Clinton's-lily, Rosy Twisted Stalk, and Canada Mayflower. Summer- and fall-flowering plants include Rugel's-ragwort, White Wood Aster, White Snakeroot, and Turk's-cap-lily.

Sick and Dying Boreal Forests

If you haven't been recently to Newfound Gap or Clingman's Dome in the Smokies or high elevations along the southern Blue Ridge Parkway, you're

Nearly all Fraser firs on Mount Mitchell, North Carolina (the highest point of the Appalachians), are dead.

in for a shock! Rather than the thriving spruce-fir forests described above, one sees large numbers, even entire stands, of Fraser Firs still standing, but dead or dying.

When the sad spectacle of dying Fraser Firs first began to be recognized in the 1970s as a serious problem, it was thought the main culprit was acid rain or, more correctly, acid deposition. More recent studies have shown that the primary cause is instead infestations by the balsam wooly adelgid (*Adelges piceae*), a small wingless insect. Acid deposition is also likely involved; it evidently weakens the trees, making them more susceptible to insect damage. Some botanists also consider global warming a factor in predisposing Fraser Firs to insect damage.

As the firs go, other species of plants and animals dependent upon them may also soon be threatened. For example, Rugel's-ragwort, an endemic of the Great Smoky Mountain National Park, is one of the shade-requiring plants found at high elevations under the firs.

CHAPTER 4

Treeless Mountain Ecosystems

Local open, treeless, or nearly so, typically grassy areas within larger forested regions were noted by early explorers of eastern North America. They also came to the attention of botanists who then, as now, found them fascinating. Such areas have been designated by such names as balds, barrens, glades, and prairies. Explanations given for the origin and distinctiveness of these treeless ecosystems have historically included some combination of rock, soil, wind, moisture, or topographic factors, possibly combined with periodic fires.

Alpine Tundra

Mountain tops are symbolic of the highest and noblest of human experiences. Mountain climbers risk their lives to feel the exhilaration of a literal "mountain top experience." Mountain tops above timberline include the alpine tundra, of special interest because of its distinctive plants.

In the Appalachians, only on the summits of Quebec's and New England's highest peaks are timberlines and alpine plants found. Included are Mt. Jacques Cartier (3,935 ft.) of the Gaspé Peninsula; Mt. Washington (6,288 ft.) and others of the Presidential Range of northern New Hampshire; Mt. Katahdin (5,267 ft.) of northcentral Maine; and Mt. Mansfield (4,393 ft.) and several smaller peaks in Vermont.

Above timberline, on and among the ubiquitous strewn rocks, are mats of lichens and mosses, various flowering herbs, and dwarfed shrubs that could easily be mistaken for herbs. The tenacity of these plants enables them to endure the thin rocky soil, below-freezing temperatures much of the year, and strong winds.

Among the adaptations that allow alpine plants to survive is their low growth habit, which helps them avoid the drying effects of wind and allows them to be insulated in winter by a blanket of snow. Leaves are thick, waxy, and tightly clustered to reduce water loss. Most alpine plants are long-lived perennials, but a few are annuals that are able to complete their life cycle within the short growing season.

Some of the more common alpine herbs are Northern Sandwort, Diapensia, Three-toothed Cinquefoil, Alpine Violet, Mountain Aster, Mountain Goldenrod, Boreal Yarrow, and Moss Campion. Also present, especially on more exposed sites, are various sedges, grasses, and rushes.

Among the shrubs, the heath family (Ericaceae) is well represented by such species as Bog Bilberry, Mountain Cranberry, Lapland Rosebay, Alpine Azalea, Moss Plant, Mountain Heather, and Snowberry. Alpine Willow grows here as a shrub.

Shale Barrens

Appalachian shale barrens are found in the middle Appalachians within a broad band from southwestern Virginia and adjacent West Virginia, north to

Shale barrens, Greenbrier County of southeastern West Virginia. Virginia Pine (*Pinus virginiana*) is in the background.

southcentral Pennsylvania. They occur on steep, south-facing slopes of hills at an elevation between 1,000 and 2,000 feet. The underlying shale, of Devonian origin, has eroded to form a distinctive, slightly acidic, brownish yellow soil. On the surface, instead of the usual organic material, is a highly mineral layer with rock flakes on the surface. Thus shale barren plants, including several endemics, must be able to tolerate poor, dry soils, as well as endure high summer temperatures.

Among the fifteen shale barren endemics (all herbs) listed by Earl L. Core in his *Vegetation of West Virginia* are Nodding Onion, Yellow Buckwheat, Leatherflower, Shale Rockcress, Shale Barren Clover, Shale Barren Evening Primrose, Mountain Pimpernel, Pursh's-bindweed, Buckley's-phlox, Barren Goldenrod, Oblong Aster, Pussytoes, and Shale Ragwort.

Other plants, not endemics but often found in shale barrens, include Prickly-pear Cactus, Wild Pink, Birdfoot Violet, Wild Stonecrop, Lime Stonecrop, Goat's-rue, Butterflyweed, Gray Beardtongue, Leonard's-skullcap, Tall Bellflower, Spiked Gayfeathers, and Shale Hawkweed.

Serpentine Barrens

These barrens are another example of a special mineral/soil type that has produced a unique flora. The vegetation of serpentine areas is often distinguished from that of more usual soils because of its stunted appearance. Also called ultramafic, such minerals, their soils, and distinctive plants are found throughout the world, especially along lines that mark the boundaries of fused geological plates (see chap. 1).

The name "serpentine" is obscured in antiquity but could come from the greenish, mottled appearance of the soil, suggestive of the markings on certain snakes. Serpentine soils are produced by the weathering of metamorphic rocks with more than 70 percent iron and magnesium compounds. Although having high concentrations of heavy metals such as iron, chromium, and nickel, they are deficient in the nutrients plants need in largest amounts: nitrogen, phosphorous, and potassium.

In the Appalachians, serpentine soils and their vegetation extend in discrete and localized sites for the entire length of the mountain chain, including serpentine barrens from northern Georgia to the Conowinga barrens of Maryland to Vermont to Mt. Albert, the high point of the Shickshocks at the tip of the Gaspé Peninsula. Mt. Albert is, in fact, a large serpentine plateau.

Mysterious Grassy Balds

"Balds" are open, treeless areas found above the spruce-fir forests, especially those of the southern Appalachians. Although they sometimes intergrade, we can recognize two kinds of mountain balds, grassy balds and heath balds.

Serpentine barren of Soldier's Delight, Maryland, contrasts with the Virginia pine forest; the wildflower is an especially hairy variety of field cerastium (*Cerastium arvense* var. *villosissimum*).

Grassy balds are prairie-like forest openings containing grasses, herbs, and sometimes a few scattered shrubs. They are located on or near ridgetops, usually at elevations of 5,000 to 6,000 feet; there are approximately 2,500 acres of grassy balds in Appalachia. Among the best known are Gregory Bald and Andrews Bald, both in the Great Smoky Mountains National Park, and Roan Mountain, actually a long high ridge, about a hundred miles to the northeast of the park on the North Carolina–Tennessee border. (For details of these sites, see appendix 2).

Grasses of balds include the native Wild Oatgrass (or Allegheny Fly-back), which is the dominant grass, along with Timothy and Kentucky Bluegrass. Among the wildflowers are Mountain Sandwort, Mountain St. John's-wort, Three-toothed Cinquefoil, Michaux's-saxifrage, Filmy Angelica, Robbin's-ragwort, and various asters.

Scattered shrubs include Mountain Rosebay, Flame Azalea, Mountain Laurel, Minnie-bush, several species of blueberries, Green Alder, and several species of hawthorn.

There is a mysterious aspect to mountain balds. What accounts for the origin and maintenance of grassy balds? Why has succession not converted them all to forests? These questions have been partly, but not entirely, answered by ecologists. It is likely that a variety of factors are responsible.

Heath Balds

Although less conspicuous than grassy balds, heath balds are more prevalent and occur in much the same general areas as do grassy balds. Heath balds of various sizes are found mostly on steep ridges with extensions onto south- or west-facing slopes. In West Virginia they are called "huckleberry plains"; in the Smokies, "laurel slicks" or "laurel hells." Though heath balds are more common in the southern Blue Ridge than farther north, one of the largest and most representative is on the top of Spruce Knob, West Virginia, at 4,840 feet the highest point in the state (see appendix 2).

As you would expect from their name, most of the shrubs of heath balds belong to the heath family, Ericaceae: Great Rhododendron, Mountain Rosebay, Flame Azalea, Mountain Laurel, Minnie-bush, Trailing Arbutus, Southern Mountain-cranberry. Other dominants include Mountain Holly and Red Elderberry.

In addition to the dominant shrubs, a variety of lichens, mosses, and clubmosses cover the thick peaty soil of southern heath balds. Flowering herbs include Trailing Arbutus, Galax, Fireweed, Hairy Sasparilla, Painted Trillium, as well as many of the herbs listed above for grassy balds. For heath balds of West Virginia, Wild Bleeding-heart, Three-toothed Cinquefoil, Bunchberry, and Gaywings are important.

Disturbed Mountain Areas

Following disturbance of an area, there is a rapid invasion by plants. These pioneer plants, many of which are considered weeds, typically produce large numbers of tiny seeds that are easily dispersed into open, sunny habitats. The seeds germinate quickly under these conditions, producing first seedlings and soon mature plants with a tolerance for the warmer, drier conditions typical of disturbed areas. They also have relatively short life cycles that permit them to flower and produce many seeds within only a few months.

Within the Appalachians, as elsewhere in eastern North America, disturbed sites are widespread. In fact, the mountains are a crazy quilt, a patchwork of landscapes, some of which are climax ecosystems but most of which are in some stage of succession. Even before humans entered the scene,

Light green heath balds contrast with the adjacent forests of the Great Smoky Mountains National Park.

glaciation, erosion, fire, windfall of trees, and other natural disturbances were common. With the arrival and increasing numbers of humans, their animals, and their tools and machines, disturbed areas have, unfortunately, become more often the rule than the exception.

The plants of disturbed areas are a mixture of native and alien (exotic) plants. Most of the latter were introduced from Europe or Asia, but some come from lowlands of eastern North America. These weedy alien plants are generally aggressive, competing with native species for light, water, and nutrients.

In Appalachia, some of the most notorious noxious weeds include Multiflora Rose, Kudzu, and in wetlands, Purple Loosestrife. Other exotics such as Butter-and-eggs and Garlic-mustard, being much less aggressive, generally behave more like native species. Before attempting to eliminate a foreign plant, "read before you weed."

The distinction between weeds and wildflowers is not always clear. Many weeds are attractive, especially when examined under a hand lens. To exclude them from consideration would be to ignore a significant portion of the plants that are now an integral part of the Appalachian flora.

Annual weeds, plants that live only a single growing season, often dominate early successional stages. Biennials undergo only vegetative reproduction the first year; the second, they produce flowers and seeds before dying. Perennials, once established, live an indefinite number of years, typically reproducing many times.

Some conspicuous annuals of disturbed areas include Field Pansy, Cow Vetch, Showy Evening Primrose, Common Morning-glory, Giant Sunflower, Common Fleabane, and Common Dayflower. Less showy annuals are Common Chickweed, Wild Mustard, Wild Buckwheat, Beggar's-ticks, Common Ragweed, and Horseweed.

Among the weedy biennials are Viper's-bugloss, Wild Teasel, Black-eyed Susan, Common Mullein, and Queen Anne's-lace.

Many weedy perennials are composites (members of the aster family): Autumn Sneezeweed, Grass-leaved Aster, Golden-aster, Canadian Goldenrod, Common Yarrow, and Orange Hawkweed. Other weedy perennials are Pokeweed, Bouncing Bet, Curly Dock, Common St. John's-wort, Musk Mallow, Passion Flower, Partridge Pea, Fireweed, Flowering Spurge, Common Milkweed, Butterflyweed, Narrow-leaved Vervain, Lyre-leaved Sage, Heal-all, Yellow Bedstraw, and Blackberry-lily.

Some weedy wildflowers of wetlands are listed in chapter 5.

In addition to the herbs (nonwoody plants) above, woody plants—shrubs, vines, and even trees—may be invasive and therefore weedy. Among the aggressive shrubs are Sweetbrier, wild roses, blackberries, New Jersey Tea, Staghorn Sumac, and Spreading Dogbane. Woody vines include Cross Vine, Trumpet Creeper, Japanese Honeysuckle, and Poison Ivy. Among Appalachian weedy trees are Fire Cherry, Common Hawthorn, Tree-of-heaven, and Princess Tree.

CHAPTER 5

Mountain Wetlands

Considering the aquatic origin of life, it is not surprising that wetlands, transitional zones between water and land, are teeming with a great variety of life. These shallow water ecosystems, with seasonally fluctuating water levels, make a major contribution to global biodiversity.

The Wetland Environment

During the nineteenth and early twentieth centuries, federal policy encouraged the "reclamation" of wetlands. As a result, their extent was reduced from an estimated 215 million acres at the time of the nation's founding to slightly less than 100 million today. Now that we recognize the ecological roles they play, wetlands are protected by federal legislation. This requires that some way of defining wetlands be agreed upon. The criteria include water at or near the surface for part of the year; hydric soils, distinctive from upland soils; and the presence of hydrophytes, plants that tolerate having "their feet wet" for much of the year. These criteria are often difficult to apply, but continued efforts to identify, study, and protect wetlands are essential to their survival.

Ecologists classify wetlands. Marshes are wetlands dominated by grasses or other herbaceous emergent plants; swamps, by trees and shrubs. Seeps, local areas supersaturated with water, often occur on hillsides below springs, especially in the southern Appalachians. Floodplains include overflow areas parallel to streams; their vegetation varies, but in the mountains it is often similar to that of swamps. Hydrophytes are sometimes found also in wet meadows, which are somewhat like marshes but are covered by water for shorter periods. Bogs, including sphagnum glades, are highly acidic peatlands in which sphagnum moss (*Sphagnum* species) is the principal contributor to peat formation.

Like all ecosystems, wetlands are dynamic. Over time, measured in centuries or decades, gradual, directional changes (i.e., aquatic succession) occur. As an example, a marsh, as it accumulates more organic matter and soil, permits the invasion of trees, converting it into a swamp. Still later, a forest will develop as the process is completed.

Boreal Bogs

To be appreciated, a bog must be experienced. Walking on the thick mat of floating peat that surrounds and partially covers the water creates a strange sensation as the ground undulates under your feet. Throughout the cool temperate world regions where they occur, bogs have been recognized as places apart from the ordinary. Scientists consider them to be unique, discrete, multidimensional ecosystems with a distinctive flora that is worthy of protection and study.

What are bogs and how do they form? Actually, there are a number of different kinds. A typical New England mountain bog is described below. Retreating glaciers at the end of the Ice Age left "kettleholes," depressions filled with water from melting ice. Invading grasses and sedges were followed by a thick mat of sphagnum moss. After centuries, a distinct acidity and high organic content was imparted by the moss to the water and the soil around the edge of the bog.

Bogs show distinct zonation. From the center outward are (1) dark tannin-laden water, (2) an open area composed of a low growth of mosses, sedges, and herbs (typically growing on floating peat), (3) a dense stand of shrubs,

Bog of Baxter State Park, Maine, has typical zonation from open water to boreal forest. Mount Katahdin is in the background.

which is encircled by (4) bog trees of the adjacent forest. As many bogs are in the process of filling in, these zones may also be thought of as successional stages.

The Fragrant Water-lily is often seen floating on the water (zone 1); the Horned Bladderwort or Common Bladderwort (both carnivorous plants) may be found beneath the surface.

In addition to mosses, other plants of the open area (zone 2) are various sedges, such as Deer's-hair Sedge, and various cotton grasses. Several carnivorous plants inhabit bogs. You may see sundews or pitcher plants. Included among the orchids are Yellow Lady-slipper, Pink Moccasin Flower, and Rose Pogonia.

Shrubs (zone 3) are mostly heaths: Great Rhododendron, Rhodora, Sheep Laurel, various cranberries and huckleberries, Bog Rosemary, and Labrador Tea.

Only two trees (zone 4) are common: Tamarack, an unusual conifer that sheds its needles in fall, and Black Spruce. Other trees are Balsam Fir, White Spruce, White Pine, and White Oak.

Sphagnum Glades

Bogs of a somewhat different type, called "sphagnum glades," occur sporadically in the middle Appalachians, south of the line of glaciation. They are similar to northern bogs in general appearance and flora but differ in their genesis. Not of glacial origin, they are formed as mountain streams are impeded by hard surface rocks. The sedimentary rocks of the Alleghenies from Pennsylvania to Virginia are well suited to the formation of these "glades," and most are found there. One of the most accessible and representative sphagnum glades is Cranberry Glades, located within the Monongahala National Forest of eastern West Virginia (see appendix 2).

As in northern bogs, sphagnum moss is also largely responsible for the acidic conditions of sphagnum glades. They, too, are characterized by members of the sedge and heath families, together with several species of bog orchids and a few carnivorous plants. The plant list above for boreal bogs is also pertinent here.

Floodplains and Other Wetlands

Wetland plants don't make arbitrary distinctions between marshes, swamps, and floodplains, or the edges of streams, ponds, and lakes. Listed below are herbs found somewhat indiscriminately in these various wetlands: Spatterdock, Marsh Marigold, Bulbous Buttercup, Nodding Smartweed, Pink Knotweed, Spotted St. John's-wort, Swamp Rose-mallow, Common Blue Violet, Marsh Blue Violet, Watercress, Oconee Bells, Whorled Loosestrife, Grass-of-

"Round" cranberry glade, Monongahala National Forest, West Virginia, has cranberries and ferns in foreground; red spruce, in background.

Parnassus, Virginia Meadow-beauty, Cross-leaved Milkwort, jewelweeds, Golden Alexanders, Water Hemlock, Soapwort Gentian, Meadow Phlox, Virginia Bluebells, False Dragonhead, Blue-eyed Mary, Water-willow, Cardinal Flower, Autumn Sneezeweed, Bog Goldenrod, Hollow Joe-Pye-weed, New York Ironweed, Golden Club, Skunk-cabbage, Common Arrowhead, Nut-grass, Fraser's-sedge, River Oats, American Bur-reed, cattails, Pickerel-weed, Yellow Star Grass, Spider Lily, Northern Blue Flag, Yellow Iris, Showy Lady-slipper, Large Purple Fringed Orchid, and Lily-leaved Twayblade.

Woody wetland plants include Pinxter Flower, Pink-shell Azalea, Dog-hobble, Steeple-bush, Ninebark, Red Chokeberry, Southern Swamp Dogwood, Red Buckeye, and Wafer-ash.

Conserving Mountain Wetlands

Appalachian wetlands are threatened. Among the threats are acid deposition and pollution from various sources including farming, forestry, and mining operations. Fortunately, the mountains are less densely populated than are adjacent lowlands, so their wetlands are correspondingly less polluted. Moreover, regulations at both the federal and state levels are intended to protect and, on occasion restore, wetlands. These laws, together with recent executive orders, have resulted in a "no net loss" policy, which should halt further losses in these critical habitats.

CHAPTER 6

Identifying Mountain Wildflowers

Carolus Linnaeus, after extensive botanizing across Europe, developed an international reputation as a botanist at the University of Uppsala in Sweden. His fame rested largely on his book *Species Plantarum* (1753), which established the modern system of naming and classifying plants.

Naming Plants

The identification of a plant may involve determining one of its common names or its single, unique scientific name. Common names are essentially nicknames and as such are often fanciful or descriptive. Consider the herbs Jack-in-the-pulpit or Lily-leaved Twayblade. Common names can be misleading; for example, Cotton-grass is actually a sedge, not a grass. Common names are not standardized; often the same species is called by several different names, even within the same area.

Botanists use scientific names to avoid confusion. A scientific name, assigned when a new species is first discovered, must conform to the basic rules of taxonomy established by Linnaeus. Each scientific name is a binomial consisting of a genus name followed by a specific epithet. The genus is capitalized but the specific epithet is not; both are italicized. The name, abbreviation, or initial that follows the binomial indicates the person who first described the species and applied that name. For example, *Sedum telephioides* Michx. was named by the famous French botanist André Michaux. Linnaeus himself established many plant names in North America as well as in Europe; these names are followed by the initial "L." for example, *Phlox divaricata* L.

What Is a Wildflower?

Wildflowers are angiosperms (flowering plants) that grow wild. Featured in this book are not only herbs (nonwoody wildflowers) but also flowering shrubs, trees, and woody vines. Most are native, but some are exotic (alien).

To exclude exotics, even those that are "weedy," would be to ignore a good percentage of the plants one sees in the mountains, especially in open or disturbed places (see chap. 4).

To identify wildflowers, both vegetative parts (roots, stems, and leaves) and reproductive parts (flowers, fruits, and seeds) may be used. In this guide leaves and flowers, especially flowers, are emphasized.

Leaves

Leaves vary greatly in size, shape, arrangement, and numerous other ways. Some leaf variations are shown in fig. 1.

Flowers

Flowers serve primarily a reproductive function. A "typical" flower consists of four kinds of parts arranged into concentric rings called whorls (see fig. 2). From the outside inward:

Sepals (collectively, calyx) are usually green, but are sometimes petal-like (white or some color other than green).
Petals (collectively, corolla) are usually the showiest part of the flower; they may be fused to form corolla tubes.
Each stamen is typically composed of an anther, which produces pollen grains, and a slender filament below, which supports the anther.
The pistil (one or more) is composed of (from top down) the stigma, style, and ovary. The ovary becomes a fruit containing one or more seeds.

Flowers vary greatly. In some, stamens are present, but pistils are absent; such a flower is said to be staminate, in contrast to one that has only pistils, which is called a pistillate flower. Flowers with both stamens and pistils are bisexual. Monoecious plants have both staminate and pistillate flowers on the same individual, whereas those with the two kinds of flowers on separate individuals are dioecious.

In many plants, flowers occur in arrangements known as inflorescences. Some of the more common ones are shown diagrammatically in fig. 2. In each case a common stalk, the peduncle, supports a cluster of flowers. In some instances, a single type of inflorescence is associated with a particular plant family. For example, each common daisylike "flower" of the aster family (Asteraceae) is actually a head consisting of numerous tiny flowers (florets).

Monocots versus Dicots

All flowering plants belong to one of two grand groups of Angiosperms: monocots (Monocotyledones) or dicots (Dicotyledones). These names are based on the number of cotyledons (one or two) possessed by the seed embryo. As seeds are often small and their cotyledons are not easily seen, it is more practical to rely instead on leaf or flower features or both to determine to which of these groups a given plant belongs.

The leaves of monocots are typically parallel-veined, whereas those of dicots are either pinnately or palmately net-veined (see fig. 1).

The flowers of monocots are 3-merous, that is, each or most whorls contain 3 (or twice 3) parts (see fig. 2). An example is a lily with 3 sepals, 3 petals, 6 stamens, and 1 pistil. The basic number for flowers of dicots is something other than 3; usually they are either 4- or 5-merous.

FIGURE 1. LEAF STRUCTURES AND VARIATION

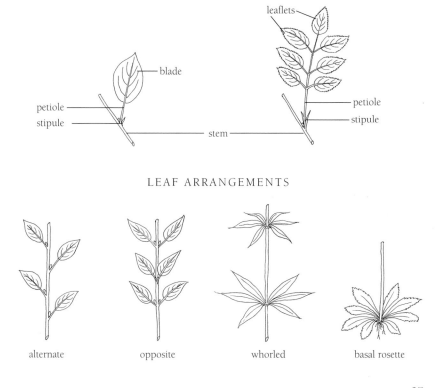

SIMPLE LEAF   COMPOUND LEAF

leaflets

blade

petiole

stipule

petiole

stipule

stem

LEAF ARRANGEMENTS

alternate   opposite   whorled   basal rosette

LEAF SHAPES

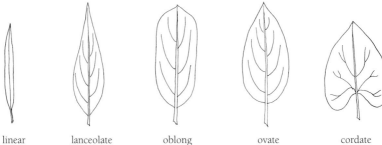

linear   lanceolate   oblong   ovate   cordate

BLADE MARGINS

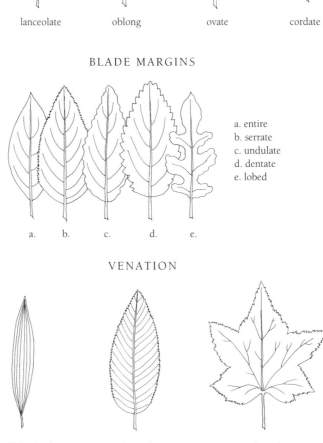

a. entire
b. serrate
c. undulate
d. dentate
e. lobed

a.   b.   c.   d.   e.

VENATION

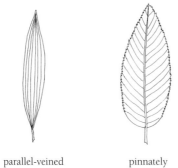

parallel-veined   pinnately
net-veined   palmately
net-veined

FIGURE 2. FLOWER STRUCTURES AND ARRANGEMENTS

TYPICAL FLOWER

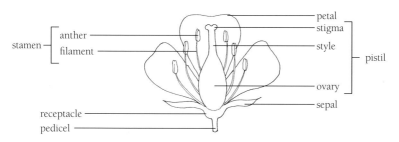

INFLORESCENCES

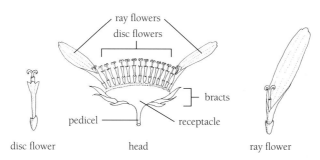

disc flower   head   ray flower

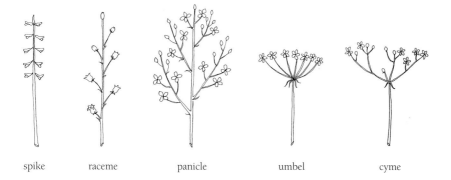

spike   raceme   panicle   umbel   cyme

Part 2

Using the Color Plates

Identifying a wildflower allows you to call it by name (common, scientific, or both), thereby opening up a wealth of information about that particular plant.

The initial step is to recognize to which of the seventeen categories your unknown plant belongs (see table of contents and "Using This Book"). The categories are based primarily on flower (or fruit) color and secondarily on the grand groups of plants known as monocots or dicots. (Within each of the categories, plants are arranged by families.)

In addition to the more than 300 species pictured and described in some detail, many related species are compared. Thus more than 800 species of Appalachian plants can be identified using this book.

"Flower" refers to the showiest part of a plant, usually sepals or petals of the flower itself, but sometimes colorful bracts (modified leaves) associated with the flowers. In the case of flower "heads," such as those of daisies, the color of the petals of the ray florets that surround the head are to be used. For bicolored flowers, the darker or more prominent color is used in the key.

Defining Terms

"Fruit" is used in the botanical sense to include berries, capsules, and other seed-containing parts of a plant.

"Woody plants" include trees, shrubs, and vines that develop hard, woody stems.

"Herbaceous plants" (herbs) are those with stems that lack woody tissues.

"Monocots" are plants with parallel-veined leaves and floral parts in 3s (example: 3 sepals, 3 petals, 6 stamens, 1 pistil).

"Dicots" are plants with net-veined leaves and floral parts in 4s or 5s (example: 4 sepals, 4 petals, 8 stamens, 1 pistil).

A rather detailed description of flowers, leaves, and terms used to describe them is given in chapter 6, "Identifying Mountain Wildflowers." Other terms are defined in appendix 1, "Glossary."

White

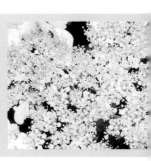

Common Arrowhead   *Sagittaria latifolia* Willd.
Water Plantain Family   Alismataceae

Arrowheads are widespread aquatic plants whose flowers have 3 white, rounded petals. The shape of the leaves varies with the species; in this one, they are 8–10 in. long and very broadly arrow-shaped. Common Arrowhead is found in shallow water along the edges of ponds and slow-moving streams from s. PA and WV southward; Jul.–Oct. Grass-leaved Arrowhead (*S. graminea*), with leaves consistent with its name, is also found throughout our area, but less frequently; Jul.–Sep.

Indians, who used the name Wapato for arrowheads, cooked and ate the tubers like potatoes.

Fraser's Sedge   *Cymophyllus fraseri* (Andr.) Mackenz. *[Carex fraseri]*
Sedge Family   Cyperaceae

This sedge, named for its discoverer, John Fraser, is showier than other species, which typically resemble grasses (see green/brown section). The straplike leaves (1–2 in. wide and 2 ft. long) and large flower clusters are distinctive. It is uncommon in the s. Appalachians except in the Great Smoky Mountains, where it is often found in deciduous woods and along streams at elevations from 1,500 to 4,000 ft.; May–Jul.

Cotton-grass (Linaigrette)   *Eriophorum spissum* Fern.
Sedge Family   Cyperaceae

Related neither to grasses nor cotton, the several Appalachian species of *Eriophorum* occupy bogs and meadows, principally in the tundra north of our area; many are circumboreal. This species, also known as Hare's-tail, grows to 1 ft. on acidic peatlands in the mountains s. to PA; Apr.–Jul.

The similar Tawny Cotton-grass (*E. virginicum*), with tan, hairy flower clusters, is found at high elevations in the s. Appalachians as well as at lower ones further north to Que.; Jul.–Sep.

Common Arrowhead
Sagittaria latifolia

Fraser's Sedge
Cymophyllus fraseri

Cotton-grass
Eriophorum spissum

Fly Poison *Amianthium muscaetoxicum* (Walt.) A. Gray
Lily Family Liliaceae

The lily family is a large and widespread group of plants, mainly of temperate regions. A typical flower includes 3 sepals, 3 petals, 6 stamens, and a long pistil with a 3-lobed stigma at its end. Nearly all are perennials that overwinter as bulbs.

The upright or bending stalks (to 4 ft.) of Fly Poison bear small, white flowers that turn from white to green to purple as they age. Basal leaves are linear and up to 2 ft. long. It is found in sandy woods and roadsides from PA southward; May–Jul.

All parts of the plant, but especially the bulbs, contain poisonous alkaloids. Colonists mixed a paste made from the bulbs with sugar to kill flies.

Bunchflower (*Melanthium virginicum*), found in bogs, looks similar but has petals that are narrower at the base; Jun., Jul.

Spanish-bayonet *Yucca filamentosa* L.
Agave Family Agavaceae

Also called Bear-grass, the tall (6–9 ft.) panicles of this perennial contain white flowers, each with 6 tepals (3 sepals, 3 petals). Primarily a Coastal Plain species, it is found sporadically in dry sandy soils of the s. Appalachians, especially in old fields and pine forests; May–Jul.

Fibers from the thick, leathery, pointed basal leaves have been used for cord. Roots contain saponins, which accounts for their use as fish poisons and in soaps and shampoos. Both the petals and fruits are edible.

Devil's-bit *Chamaelirium luteum* (L.) A. Gray
Lily Family Liliaceae

Also called Fairy-wand, this perennial has spikes with tiny greenish or white flowers that turn yellowish with age. Basal leaves form a rosette. It is rather common in rich woods from NY and w. MA southward; Mar.–May.

Turkey Beard (*Xerophyllum asphodeloides*) is a similar but less common wildflower found sporadically along the Blue Ridge Parkway, especially n. of Asheville. It has tall (5 ft.) flowering stalks with spikes of small white flowers and narrow grasslike leaves; Jun.

Colic-root *Aletris farinosa* L.
Lily Family Liliaceae

A tall (2–3 ft.) stalk bears the small urn-shaped flowers, each swollen at the base. Long, sharp-pointed, lanceolate leaves form a basal rosette (see fig. 17). It is found in sandy or peaty soils from s. New Eng. southward; Apr.–Jun.

A root decoction has been used to treat indigestion, colic, and other ailments. It contains diosgenin, known to be anti-inflammatory.

Fly Poison
Amianthium muscaetoxicum

Spanish-bayonet
Yucca filamentosa

Devil's-bit
Chamaelirium luteum

Colic-root
Aletris farinosa

False-garlic   *Nothoscordum bivalve* (L.) Britt.
Lily Family   Liliaceae

Once believed to be more closely related to onions and garlic (*Allium* spe-cies), this plant lacks the strong odor of those plants, thus False-garlic. The white or cream flowers, each ½ in. wide, are arranged in clusters of 3–10 at the top of a ft.-tall stem. Basal leaves are linear. It occurs on rocky outcrops and moist prairies from VA s. to TN and SC; Mar.–May.

Canada Mayflower (Maguet)   *Maianthemum canadense* Desf.
Lily Family   Liliaceae

As this small (3–6 in.) perennial grows from spreading rhizomes, it often forms large colonies (as seen here). Each aboveground stem bears 2–3 cor-date leaves and a fragrant raceme. Flowers are starlike with 4 petals; when mature, berries are coral. Canada Mayflower is often locally common, espe-cially in high-elevation forests throughout the Appalachians. Also called Wild Lily-of-the-valley, it is sometimes cultivated as a ground cover; May, Jun.; fruits, Jun.–Sep.

This is the only species of its genus, but several others of the lily family have a similar appearance. Lily-of-the-valley (*Convallaria majalis*) has bell-like white flowers; Apr., May. Three-leaved False Solomon's-seal (*Smilacina trifolia*), a bog plant of the n. Appalachians, has lanceolate leaves that sheathe the stems and flowers with 6 petals; the raceme also is less compact; May, Jun.

Spider-lily   *Hymenocallis caroliniana* (L.) Herb. [*H. occidentalis*]
Lily Family   Liliaceae

This striking perennial, which grows from a large bulb, has linear basal leaves. The flower stalk, which may be 3–4 ft. tall bears several (3–7) flow-ers in a large umbel (see fig. 18). From the long slender corolla tube extend 6 narrow tepals, which are connected by the round membranous crown; at-tached to it are the 6 long stamens. Spider-lily is principally a plant of low-lands west, south, and east of the s. Appalachians, but it is found, though infrequently, in low-elevation swamps and along stream banks in the moun-tains of NC, SC, and TN; Jun.–Aug.

This is the only spider lily of our area.

False-garlic
Nothoscordum bivalve

Canada Mayflower
Maianthemum canadense

Spider-lily
Hymenocallis caroliniana

Large-flowered Trillium (Trille Grandiflore)

Trillium grandiflorum
(Michx.) Salisb.

Lily Family   Liliaceae

Trilliums have their leaves, sepals, and petals all in whorls of 3. The dozen or so Appalachian species of trillium have petals that vary from white to pink, yellow, or red. Trilliums fall into two categories: stalked, which bear flowers on a stalk above the leaves; and sessile, whose flowers lack stalks.

Large-flowered Trillium (the floral emblem of Ontario) is about 18 in. tall and has stalked flowers. The petals, white when they first open, turn pink as they age. It inhabits rich woods from s. Que. to n. GA; Apr.–Jun.

Two other white, stalked trilliums have more restricted ranges. Sweet White Trillium (*T. simile*), limited to the Great Smoky Mountains region, is quite similar but has a dark purple ovary surrounded by "gaping" (3-D) petals; Apr., May. Snow Trillium (*T. nivale*) is a tiny (petals less than $1\frac{1}{2}$ in. long), primarily midwestern species sometimes seen in WV and PA; Mar., Apr.

White Wake-robin

Trillium erectum L.

Lily Family   Liliaceae

This is one of the color variants of Red Trillium or Wake-robin (see red section); besides red, others may be cream, yellow, or pink. The white form here can be distinguished from Large-flowered Trillium (*above*) by its narrower petals and dark pistils; Apr., May.

Nodding Trillium (Trille Penche)

Trillium cernuum L.

Lily Family   Liliaceae

This trillium has nodding, sweet-scented flowers with reflexed white or pink petals and pink anthers. It is a Great Lakes and n. Appalachian species found s. (rarely) to VA; Apr., May.

South of the range of Nodding Trillium, especially in oak woods, is found Catesby's Trillium (*T. catesbaei*). Its nodding white flowers are similar but have bright yellow anthers; Apr., May.

Large-flowered Trillium
Trillium grandiflorum

White Wake-robin
Trillium erectum

Nodding Trillium
Trillium cernuum

Solomon's-seal   *Polygonatum biflorum* (Walt.) Ell.
Lily Family   Liliaceae

This 3–4 ft. tall perennial with an arching stem has smooth alternate leaves and 8–12 yellow flowers arranged in pairs (*biflorum*) at their nodes. Berries that follow are brownish. It occurs in a wide range of dry to moist habitats throughout most of e. N. Amer. from New Eng. southward; May, Jun.; fruits, Jun.–Oct.

A root tea was widely used by American Indians and white settlers for a variety of digestive and joint disorders.

Also called Solomon's-seal, *P. pubescens* has hairs along the veins on the underside of its leaves; Apr.–Jun.

Solomon's-plume (Smilacine a Grappes)   *Smilacina racemosa* (L.) Desf.
Lily Family   Liliaceae

In contrast to Solomon's-seal (*above*), this species has tiny, white starlike flowers in a terminal panicle. The fruit is a red berry. The stem is somewhat zigzag. Sometimes called False Solomon's-seal, it is found in much the same habitats and range as Solomon's-seal; May–Jul.; fruits, Jun.–Oct.

Indians used a leaf tea as a contraceptive and for coughs; root tea, as a stomach tonic and purgative. The young shoots can be eaten like asparagus or added to salads.

Starry False Solomon's-seal (*S. stellata*) is similar but has larger flowers and black berries; Apr.–Jun.; fruits, Jun.–Oct.

Smilax   *Smilax ecirrhata* (Engelm.) S. Wats. [*S. biltmoreana*]
Catbrier Family   Smilacaceae

Smilax species are generally woody or herbaceous vines that climb on fences and form entanglements, often by means of tendrils. Many have prickles, accounting for such names as Sawbrier, Greenbrier, and Catbrier. Small, bell-like flowers are suggestive of those of the lily family (to which *Smilax* has traditionally been assigned).

This herbaceous species has neither tendrils nor prickles. The undersides of the leaves are hairless but covered with a white bloom that is easily rubbed off. It is infrequently found in deciduous woods at low elevations in the s. Appalachians (more commonly in the Midwest); May, Jun.

Smilax hugeri is similar but has leaves that are hairy beneath; Mar.–Apr. Some *Smilax* species, especially *S. herbacea*, are called Carrion-flower; their odor is like that of decaying meat, and they are pollinated by carrion flies. It has round umbels, each composed of several dozen small white flowers; May, Jun.

Various *Smilax* species were used medicinally by Native Americans: root tea for rheumatism and stomach troubles and to expel the afterbirth.

Solomon's-seal
Polygonatum biflorum

Solomon's-plume
Smilacina racemosa

Smilax
Smilax ecirrhata

Wild Yam   *Dioscorea villosa* L.
Yam Family   Dioscoreaceae

Belonging to a mainly tropical family of twining vines, Wild Yam is characterized by shiny cordate leaves arranged alternately or in whorls. The small white flowers are in panicles attached at the base of the leaves. It is found in wet woods and swamps from s. New Eng. s. to TN; May–Aug.

American Indians used the roots of this plant to make a tea used in the treatment of a large number of ailments including colic, thus sometimes called Colic-root.

Chinese Yam (*D. batatas*), or Cinnamon-vine, is a somewhat similar vine of drier waste places in the Appalachians; it has opposite leaves, each with squared bases; Jun.–Aug.

Hatpins   *Eriocaulon decangulare* L.
Pipewort Family   Eriocaulaceae

Eriocaulon species are distinctive wetland plants found mainly on the southeastern Coastal Plain. This species, an uncommon inhabitant of the Carolina mountains, bears compact clusters of white flowers on stems 2–3 ft. tall; Jun.–Oct.

Also called Hatpins, *E. compressum* is a smaller plant with less compact, more flattened flower heads. It also is found in the mountains of NC and SC; Jun.–Oct. White-button (*E. aquaticum*) is a shallow-water plant more common in the n. Apps but found sporadically in the mountains to NC; despite its common name, its flowers are dark gray to black; Jul.–Oct.

Slender Ladies' Tresses   *Spiranthes lacera* (Raf.) Raf.
Nodding Ladies' Tresses   *S. cernua* (L.) Rich.
Orchid Family   Orchidaceae

The approximately 10 Appalachian species of *Spiranthes* are recognized by the spiral arrangement of the tiny white or greenish flowers; they flower principally in late summer or fall. Leaves usually wither before flowering time.

In Slender Ladies' Tresses (*left*), the slender flower stalk is 20–30 in. tall, and the flowers are arranged in a single spiral. The basal leaves are usually withered by flowering time. Look for these plants in open fields and woods at low to middle elevations in the mountains throughout our area; Jun.–Oct.

The white flowers of Nodding Ladies' Tresses (*right*) are more compact and each flower nods (bends downward) about 45 degrees from the horizontal; Aug., Sep.

Rattlesnake Plantain (*Goodyera pubescens*) has whitish flowers arranged in spikes similar to those of *Spiranthes*, but it has conspicuous white-veined basal leaves; Jul., Aug.

Wild Yam
Dioscorea villosa

Hatpins
Eriocaulon decangulare

Slender Ladies' Tresses
Spiranthes lacera

Nodding Ladies' Tresses
Spiranthes cernua

Fragrant Water-lily (Lis d'Eau)   *Nymphaea odorata* Aiton
Water-lily Family   Nymphaeaceae

Water-lilies are showy plants of shallow water; their flowers and leaves are attached by stems to roots that anchor them in the soil. In this species, leaves are large, round, and notched at the base. Flowers are 5 in. across, white (or red in form *rubra*), and fragrant. It is restricted to ponds, lakes, and slow-moving streams from Que. to TN and SC; Jun.–Sep.

Native Americans poulticed the roots to reduce inflammations; roots were also used to make tea for coughs and tuberculosis.

Doll's-eyes (Actée à Gros Pédicelles)   *Actaea alba* (L.) Miller
                                                                          [A. pachypoda]
Buttercup Family   Ranunculaceae

Also called White Baneberry, this 1- to 2-ft.-tall plant has racemes of white flowers followed by clusters of shiny white berries. Leaves are divided into several, often 7, leaflets with dentate margins. Look for it in rich woods and thickets from southern Que. s. to GA and AL; Apr., May; fruits, Jul.–Oct.

Red Baneberry (*A. rubra*), or Poison de Coulevre, is a somewhat more northern (also high elevations in s. Apps) species. Its flowers are also white but are in a more rounded cluster. Berries are red; May–Jul; fruits, Jul.–Oct.

The berries of both species are considered poisonous, although those of the European Baneberry (*A. spicata*) are better documented as being toxic.

Canada Anemone   *Anemone canadensis* L.
Buttercup Family   Ranunculaceae

Anemones are wildflowers of woodlands or meadows; they have white petal-like sepals (usually 5). This species, which grows 1 to 2 ft. tall, is typical. Note the rather large (2 in. across) flowers and the sessile sharply lobed leaves attached below. Primarily a plant of the n. Apps, it is found in meadows and thickets from the Gaspé to VA, but is much less common in the southern portion of its range; May–Jul.

Wood Anemone (*A. quinquefolia*) is a smaller, more delicate plant; its leaves are more deeply cut, giving the appearance of 3 or 5 leaflets. It is more common in the s. Apps than *A. canadensis*; Apr.–Jun.

Also often seen from New Eng. southward is Thimbleweed (*A. virginiana*). It produces a small flower with greenish petal-like sepals on a tall (2–3 ft.) stalk; its cylindrical fruiting head accounts for its common name; Jun.–Aug.

Fragrant Water-lily
Nymphaea odorata

Doll's-eyes
Actaea alba

Canada Anemone
Anemone canadensis

False Bugbane   *Trautvetteria caroliniensis* (Walter) Vail
Buttercup Family   Ranunculaceae

Also called Tassel-rue, this tall (2–4 ft.) perennial has divided leaves that are wider than long. Having no petals and sepals that soon fall off, the flower's most conspicuous feature is the numerous white stamens seen here. It is found in moist woods and along creek banks. More common w. of our area, it is fairly frequent from PA s. to GA; Jun.–Aug.

The genus, consisting of only this one species native to Japan, is named in honor of the Russian botanist Ernst R. von Trautvetter (1809–89).

Sharp-lobed Hepatica   *Hepatica acutiloba* DC.
Buttercup Family   Ranunculaceae

Hepaticas are low plants (2–5 in.) with 6–10 petal-like sepals of white or tints of blue, pink, or lavender. Under the sepals are 3 bracts (modified leaves). Note the sharply pointed leaf lobes. It is common in upland woods throughout the Apps; Mar., Apr.

Very similar, but with more rounded leaf lobes, Round-lobed Hepatica (*H. americana*) sometimes hybridizes with *H. acutiloba;* Mar., Apr.

Conforming to the outmoded "doctrine of signatures," hepaticas have been used for the treatment of liver ailments, but there is no evidence of their effectiveness.

Umbrella-leaf   *Diphylleia cymosa* Michx.
Barberry Family   Berberidaceae

This uncommon mountain plant is easily recognized by its large, deeply cleft leaves and small white flowers, each with 6 petals arranged into cymes. It grows to 1 ft. in height. Umbrella-leaf, restricted to wet soils of deciduous forests, is found especially along streams from VA to GA; May–Aug.

Cherokees used a root tea as a diuretic and antiseptic. Chinese Umbrella-leaf (*D. sinensis*), known to contain the anticancer compound podophyllo-toxin, is widely used in traditional Chinese medicine.

False Bugbane
Trautvetteria caroliniensis

Sharp-lobed Hepatica
Hepatica acutiloba

Umbrella-leaf
Diphylleia cymosa

May-apple (Pomme de Mai)   *Podophyllum peltatum* L.
Barberry Family   Berberidaceae

The leaves of May-apple are somewhat like those of Umbrella-leaf (*above*), but the 1–2 in. flowers, with their waxy petals, are quite different. It is quite common in forest openings and edges throughout most of e. N. Amer.; Apr., May; fruits, May, Jun.

The pencil-thin rhizomes have a strong purgative effect; they also contain anticancer substances and were used for this purpose by the Indians, as they also are in modern medicine (small-cell lung and testicular cancers). Ripe fruits can be eaten or used to make a flavorful jelly, but unripe fruits and other parts are poisonous.

Twin-leaf   *Jeffersonia diphylla* (L.) Pers.
Barberry Family   Berberidaceae

The unusual leaves, each with 2 half-ovate lobes, account for its common name, as well as the specific epithet. Borne on separate stalks, flowers are about 1 in. across. It is found in rich woods, especially on calcareous soil, from w. NY s. to TN and NC, but is not common; Apr., May.

This plant, named for Thomas Jefferson, was used by Native Americans to treat a wide variety of ailments.

Bloodroot (Sang-Dragon)   *Sanguinaria canadensis*
Poppy Family   Papaveraceae

Like Twin-leaf (*above*), this plant is a perennial nearly 1 ft. tall. Note the difference in leaf shape. It is found, but not commonly, in rich woods from s. Que. s. to TN; Mar.–May.

The thick horizontal rhizome contains bright orange red juice. This juice was widely used by American Indians to decorate the skin and to treat a variety of ailments; like a number of other plants used for dye purposes, it was called "Puccoon." Bloodroot contains sanguinarine, which has considerable potential for use in modern medicine but is highly toxic. Do not experiment! Its most important current commercial use is as a toothpaste additive and in other oral care products; Mar.–May.

May-apple
Podophyllum peltatum

Twin-leaf
Jeffersonia diphylla

Bloodroot
Sanguinaria canadensis

Dutchman's-breeches (Dicentra à Capuchon)

Dicentra cucullaria
(L.) Bernh.

Fumitory Family

Fumariaceae

Dicentra species have highly dissected fernlike leaves and distinctively shaped flowers. In Dutchman's-breeches, each ½-in.-long inverted flower has two inflated spurs that give it a "pantaloons" appearance, thus its common name. Its usual habitat is moist, rich woods, where it often occurs in large masses. It can be seen throughout the Apps from the Gaspé Peninsula s. to GA; Apr., May.

Although potentially poisonous, a root tea of Dutchman's-breeches has been used in folk and Indian medicine as a diuretic and to promote sweating.

Squirrel-corn (Dicentra du Canada)

Dicentra canadensis
(Goldie) Walp.

Fumitory Family

Fumariaceae

In comparison with *D. cucullaria (above)*, Squirrel-corn has flowers that are fragrant and somewhat heart-shaped. Its name comes from the roots, which resemble small grains of yellow corn. It is found in much the same habitat as Dutchman's-breeches and has a similar geographical range; Apr., May.

The flowers of both *Dicentra* species are sometimes light pink rather than white.

Northern Sandwort (Sabline)

Arenaria marcescens Fern.

Pink Family

Caryophyllaceae

"Pink" refers not to flower color but rather to the characteristic notched petals, appearing as if they had been trimmed with pinking shears.

Sandworts are small plants that form mats on sandy soils in open, sometimes wet sites; a few are alpine. Northern Sandwort is a low tufted plant with many small narrow opposite leaves on each stem. Note the 5 deeply cleft petals; each flower is about ½ in. across. This uncommon plant is found above treeline on Mt. Albert (Que.) and in Vermont's Green Mountains; Jun.–Aug.

The more widespread Mountain Sandwort (*A. groenlandica*) has smaller flowers with uncleft white petals. It is common on mountain summits from Que. to GA (also to the ME coast); Apr.–Jun.

Dutchman's-breeches
Dicentra cucullaria

Squirrel-corn
Dicentra canadensis

Northern Sandwort
Arenaria marcescens

Star Chickweed *Stellaria pubera* Michx.
Pink Family Caryophyllaceae

Chickweeds have 5 deeply cleft petals, giving the appearance of 10 petals. In this species, the sessile (or nearly so) leaves are relatively broad, and the flowers (1 in. across) have petals longer than the sepals. It is a common woodland species of the Apps from PA southward; Mar.–May.

Two other common chickweeds, both weedy aliens of open areas, are Common Chickweed (*S. media*), which has shorter leaves and petals shorter than the sepals, Apr.–Nov.; and Lesser Stitchwort (*S. graminea*), with very narrow leaves and much smaller flowers, May–Jul. Common Chickweed has been used in folk medicine to prepare a tea taken internally for coughs and externally for itching.

The native Mountain Stitchwort (*S. calycantha*) is a circumboreal species of high mountain stream banks. It is an inconspicuous, highly branched plant with tiny flowers and leaves wider than those of *S. graminea*; May–Sep.

Canada Violet (Violette) *Viola canadensis* L.
Violet Family Violaceae

Canada Violet is a "stemmed" violet (leaves and flowers on same stalk) with lanceolate leaves and flowers on short stalks. The white petals are yellow at their bases. Stems are purplish, and there are purplish tinges on the back of the petals (seen here in bud at right). Entire petals generally turn blue or purple as they age. It is common at low to middle elevations of northern deciduous forests, becoming less frequent southward to AL and GA; Apr.–Jul.

Pale Violet (*V. striata*) has petals that lack the yellow and purple markings noted above. Also, it has longer flower stalks and prominent stipules; Apr.–Jun.

A tea made from the roots of various violet species has been used by Native Americans for pain in the pelvic region. It was poulticed for boils and skin wounds.

Sweet White Violet *Viola blanda* Willd.
Violet Family Violaceae

This fragrant "stemless" violet (leaves and flowers on separate stalks) has red stems and purple stripes on the lower petals. The cordate leaves are deeply lobed. It occupies cool, moist sites in evergreen forests from Que. s. to GA; Apr., May.

Northern White Violet (*V. pallens*) is a stemless violet with smaller, nearly round leaves. Other stemless violets are named for their leaves: Large-leaved Violet (*V. incognita*); Lance-leaved Violet (*V. lanceolata*); and Kidney-leaved Violet (*V. renifolia*); Apr., May (all).

Star Chickweed
Stellaria pubera

Canada Violet
Viola canadensis

Sweet White Violet
Viola blanda

Watercress (Cresson de Fontaine) *Rorippa nasturtium-aquaticum*
(L.) Hayek *[Nasturtium officinale]*
Mustard Family   Brassicaceae

"Cruciferae," the traditional name for the mustard family, recognizes the 4 petals that form a cross. The fruit, called a silique, is an elongated seed pod divided by a partition into 2 seed chambers.

Watercress, often rooted in shallow water of freshwater streams, forms dense mats that creep onto the bank. Leaves are divided into 3–9 smooth, rounded leaflets. Small white flowers are arranged in panicles. A European plant originally cultivated there, it has escaped and is found in temperate regions throughout the world; Mar.–Nov.

The leaves and stems of Watercress are excellent in salads but should be washed carefully before eating.

Garlic-mustard *Alliaria petiolata* (Bieb.) Cavara & Grande *[A. officinalis]*
Mustard Family   Brassicaceae

Note the simple alternate leaves of this 1- to 3-ft.-tall biennial. The white flowers are arranged in racemes. Unlike most other alien species found in open, disturbed places, Garlic-mustard thrives in semishaded woods; it is also less aggressive than most other weedy plants. It is found throughout our area; Apr.–Jun.

The leaves, which have a mild garlic odor, can be cooked as a potherb.

Five-parted Toothwort *Cardamine concatenata* (Michx.)
O. Schwarz *[Dentaria laciniata]*
Mustard Family   Brassicaceae

This is perhaps the most common of the toothworts, low early-flowering plants with racemes of 4-petaled white to pink petals. This species, about a foot tall, is characterized by deeply incised, whorled leaves. It occurs in various habitats throughout our area but is most often found in rich woods; Mar.–May.

Other Appalachian toothworts with very similar flowers include Broad-leaved Toothwort (*C. diphylla*), with paired leaves divided into 3 broader leaflets, Apr., May; and Dissected Toothwort (*C. multifida*), with leaves more finely divided than those of *C. laciniata*, Apr., May.

Both "toothwort" and *Dentaria* refer to the traditional practice of chewing the pungent roots as a toothache remedy. They also add a spicy taste to salads.

Watercress
Rorippa nasturtium-aquaticum

Garlic-mustard
Alliaria petiolata

Five-parted Toothwort
Cardamine concatenata

Early Saxifrage　　　　　　　　　*Saxifraga virginiensis* Michx.
Saxifrage Family　　　　　　　　　　　　　　　Saxifragaceae

Panicles of small ($\frac{1}{4}$ in. across) white flowers are borne on ft.-long stalks. Like most other members of this family, each flower has 5 sepals, 5 petals, and 10 stamens. Ovate leaves form a rosette at the base of the flower stalk. It is a common plant of dry, rocky woods and ledges from VA and WV southward; Apr.–Jun.

Michaux's Saxifrage　　　　　　　　*Saxifraga michauxii* Britt.
Saxifrage Family　　　　　　　　　　　　　　　Saxifragaceae

The rosette of prominently toothed, lanceolate leaves is the best feature for identifying this saxifrage. The small white flowers are borne on the branching stem, which reaches 6–24 in. tall. It is a common wildflower of moist rocky sites, especially seeps, from VA s. to GA; Jun.–Oct.

The names recognize its discoverer, the French botanist Andre Michaux (1746–1802).

The similar Mountain Lettuce (*S. micranthidifolia*) has long (to 1 ft.) oblong leaves; May, Jun.

Two other saxifrages are alpine plants of the n. Apps: Gaspésian Saxifrage (*S. gaspensis*), with ovate leaves, Jul.; and White Mountain Saxifrage (*S. aizoon*), with narrow leathery leaves in a rosette, Jun.–Aug.

Brook-saxifrage　　　　　　　　　*Boykinia aconitifolia* Nutt.
Saxifrage Family　　　　　　　　　　　　　　　Saxifragaceae

This close saxifrage relative, which grows to 2 ft. tall, has flowers similar to those of saxifrages. Note the sharp-lobed leaves that resemble those of monkshood (*Aconitum* species), thus *aconitifolia*. It occupies moist woods from VA and WV s. to GA and AL; Jun., Jul.

Early Saxifrage
Saxifraga virginiensis

Michaux's Saxifrage
Saxifraga michauxii

Brook-saxifrage
Boykinia aconitifolia

Grass-of-Parnassus   *Parnassia asarifolia* Vent.
Saxifrage Family   Saxifragaceae

The distinctive white-petaled flowers, with their prominent green veins, are typical of *Parnassia* species. The 1- to 1½-ft.-tall plant, with its reniform leaves attached oppositely on the flower stalk, grows in seeps and swamp edges from VA and WV southward; Aug.–Oct.

The less common Big-leaved Grass-of-Parnassus (*P. grandifolia*) has larger, elongated leaves; Sep., Oct. More widespread species (but lacking recognized common names) are *P. glauca* (single, small, spadelike leaf attached to flower stalk; n. and middle Apps), Jul.–Oct.; and *P. kotzebuei* (deltoid leaves; tundra of Greenland to Gaspé Peninsula), Jul., Aug.

False Goatsbeard   *Astilbe biternata* (Vent.) Britt.
Saxifrage Family   Saxifragaceae

This tall (3–5 ft.) perennial has large leaves that are quite variable but are typically divided into several leaflets, the terminal one 3-lobed or nearly so (as seen here). White or cream flowers are arranged in huge compound panicles, which often bend over after a heavy rain. Having male and female flowers on separate plants, it is dioecious; the male plant with its more conspicuous flowers is pictured here. False Goatsbeard is found in moist woods at low to medium elevations from VA and WV southward; May, Jun.

The more common and more widely distributed (Catskills southward) Goat's-beard (*Aruncus dioicus*) of the rose family is strikingly similar in general appearance. Goatsbeard can be distinguished from False Goatsbeard by its 20 (vs. 10) stamens and 3–4 (vs. 2) seed pods per flower. When these features cannot be determined, a rule of thumb is that Goat's-beard has unlobed (vs. 3-lobed) terminal leaflets; May, Jun.

Foamflower   *Tiarella cordifolia* L.
Saxifrage Family   Saxifragaceae

A foot or less in height, this plant bears racemes of delicate white flowers (note the long stamens) on stalks separate from those that bear the simple, slightly lobed leaves. It is common in rich woodlands from New Bruns. and New Eng. southward; Apr.–Jul.

Bishop's-cap (Mitrelle à Deux Feuilles)   *Mitella diphylla* L.
Saxifrage Family   Saxifragaceae

The pair of leaves attached to the flower stalk, together with the conspicuous inflorescence with smaller, less showy flowers, distinguishes Bishop's-cap from Foamflower (*left*). If examined closely, the intricately fringed petals can be seen. Also called Miterwort, it is found in rich woods from s. Que. southward; May–Aug.

Grass-of-Parnassus
Parnassia asarifolia

False Goatsbeard
Astilbe biternata

Foamflower
Tiarella cordifolia

Bishop's-cap
Mitella diphylla

Blue Ridge Phacelia   *Phacelia fimbriata* Michx.
Waterleaf Family   Hydrophyllaceae

"Waterleaf" refers to the mottled leaves, which may appear as if marked by water droplets. Phacelias, like other members of the waterleaf family, have flowers with the calyx and corolla 5-lobed. The 5 stamens generally extend beyond the corolla. Other phacelias are pictured in the blue/purple section.

Also called Fringed Phacelia, *P. fimbriata* is a low-growing (3–5 in. tall) plant with flowers ½ in. across. It is our only phacelia with white, fringed petals. It commonly forms large solid masses in cove hardwood forests of the s. Apps; Apr.–Jun.

Miami-mist (*P. purshii*) has fringed blue flowers; Apr.–Jun. Small-flowered Phacelia (*P. dubia*) has fringeless white (or pale blue) flowers; Apr., May.

Spatulate-leaved Sundew (Rossolis)   *Drosera intermedia* Hayne
Sundew Family   Droseraceae

A sticky fluid secreted by specialized leaves of sundews attracts and holds insects, which are then digested. Note that the small (½ in. long) leaves of this species are ovate; the small white flowers are borne on a 4–8 in. stalk. It grows in bogs throughout most of the Appalachians but is more common in the Great Lakes region and along the Atlantic Coast; Jul., Aug.

Two other Appalachian sundews with white flowers: Round-leaved Sundew (*D. rotundifolia*), with a rosette of round leaves, Jun.–Sep.; and Slender-leaved Sundew (*D. linearis*), with erect elongated leaves, Jul., Aug. Pink Sundew (*D. capillaris*), of the Cumberland Plateau, has wedge-shaped leaves and pink flowers; May–Aug.

Mountain White Potentilla   *Potentilla tridentata* Sol.
Rose Family   Rosaceae

This creeping perennial has a mat of leaves, with each leaf divided into 3 leaflets and each leaflet bearing 3 rounded teeth at the tip. The tiny (less than ½ in. across) flowers are borne on thin wiry stems 10–12 in. long. It is primarily a tundra plant of Canada and Greenland with a range extension southward along high mountain tops to GA; also seen on rocky ledges along the New England coast; May–Oct.

Other cinquefoils (*Potentilla* species) are featured in the yellow section.

Blue Ridge Phacelia
Phacelia fimbriata

Spatulate-leaved Sundew
Drosera intermedia

Mountain White Potentilla
Potentilla tridentata

Bowman's-root *Porteranthus trifoliatus* (L.) Britt. *[Gillenia trifoliata]*
Rose Family Rosaceae

Bowman's-root is a smooth, 2- to 3-ft.-tall herbaceous perennial. It has leaves divided into 3 sharply pointed leaflets and flowers with 5 narrow petals. It is quite common in rich woods and shaded banks from NY southward; May–Jul.

American Ipecac (*P. stipulatus*) is very similar, but it has 2 large stipules at the base of each leaf, giving the effect of 5 leaflets; Jun., Jul.

Because of their widespread medicinal uses, both *Porteranthus* species are called Indian Physic. A leaf tea has been used to treat colds, asthma, and indigestion; a poultice, insect stings and rheumatism.

Common Strawberry (Fraisier) *Fragaria virginiana* Duchesne
Rose Family Rosaceae

Like the cultivated strawberry, this native plant is a sprawling herb that spreads by runners. Note the large teeth on the rounded ends of the 3 leaflets. Flower petals are about ¼ in. long. It is found in open, moist sites throughout the Apps; Apr.–Jul; fruits, May–Jul.

Wild strawberries, especially those of *F. virginiana*, are smaller than those of cultivated varieties but are considered tastier.

Indian Strawberry (*Duchesnea indica*), an Asian species, often occurs in waste places. It has leaves like those of *Fragaria*, but its flowers are yellow and its smaller red fruits are inedible; Apr.–Aug.; fruits, May–Sep.

Bunchberry (Quatre-Temps) *Cornus canadensis* L.
Dogwood Family Cornaceae

Cornus species of N. Amer. fit into three groups: herbs, bracted trees, and bractless trees. (Examples of tree species of *Cornus* are featured in the "Woody Dicots with White Flowers" section). Bunchberry is a 4- to 8-in.-tall herb with a cluster of small greenish flowers surrounded by 4 white bracts. Note also the 6 whorled leaves below the bracts. Fruits are round scarlet berries. A plant of moist boreal forests, it is common in Greenland, Canada, and Alaska but also ranges in the mountains as far south as VA and WV; May–Jul.; fruits, Jul., Aug.

Swedish Bunchberry (*C. suecica*) has several pairs of opposite leaves and dark purple flowers. Of Eurasian origin, it is locally abundant along the St. Lawrence R. in Que.; Jun., Jul.; fruits, Aug., Sep.

Bowman's-root
Porteranthus trifoliatus

Common Strawberry
Fragaria virginiana

Bunchberry
Cornus canadensis

Spikenard (Aralia à Grappes)
Ginseng Family

Aralia racemosa L.
Araliaceae

Aralia species are small, generally herbaceous, woodland plants with large aromatic roots, compound leaves, and tiny white flowers in clusters. Spikenard, which grows to 4 ft., has leaves divided into 5–21 dentate leaflets and flowers in small umbels arranged into racemes. It is found in moist woods from the Gaspé to GA; Jun.–Aug.

Wild Sasparilla (*A. nudicaulis*) is a smaller (2 ft.) plant with 3–5 leaflets and flowers in a single small umbel below the leaves; May, Jun. Hairy Sasparilla (*A. hispida*) is similar to Wild Sasparilla but has larger umbels (³/₄–1 in.) and stiff bristles on the stem; Jun., Jul.

Shinleaf
Shinleaf Family

Pyrola elliptica Nutt.
Pyrolaceae

Pyrolas are native herbs of cool northern and mountain forests. Above rounded basal leaves is a raceme of white flowers; in all species the flowers are similar to those seen here.

Shinleaf is a 5- to 10-in.-tall plant with elliptical leaves and petioles that are shorter than the blades. Note the long, curved pistil of each flower. It is found across Canada to New Eng. and s. on high mountains to VA and WV; Jun.–Aug.

Other pyrolas include One-sided Pyrola (*P. secunda*), with flowers displaced to one side of the stem, Jun.–Aug.; Green-flowered Pyrola (*P. virens*), with smaller leaves and green-veined flowers, Jun.–Aug.; and Round-leaved Pyrola (*P. rotundifolia*), a somewhat larger plant with leaves that are more rounded and leathery and that have petioles longer than the leaf blades, Jul., Aug.

Indian-pipe
Indian-pipe Family

Monotropa uniflora L.
Monotropaceae

Monotropa species are saprophytic; lacking chlorophyll, their roots obtain nutrients from decomposing organic matter in the soil made possible by a mycorrhizal relationship with soil fungi. Indian-pipe is a small plant (3–8 in. tall) with a translucent stem covered with scalelike leaves. The single nodding flower turns from white or pink to black as it ages. It is found in thick, woodland soils throughout our area; Jun.-Sep.

Pinesap (*M. hypopithys*) is a similar and also widespread plant, but it has yellowish to reddish stems and flowers in racemes; May-Oct.

Sweet Pinesap (*Monotropsis odorata*) is a smaller (2 to 4 in.), sweet-smelling rose to purple plant of the same family. It is restricted to pine woods of the s. Apps; Feb.-Apr.

Spikenard
Aralia racemosa

Shinleaf
Pyrola elliptica

Indian-pipe
Monotropa uniflora

Diapensia   *Diapensia lapponica* L.
Diapensia Family   Diapensiaceae

This family is related to the heath family (Ericaceae) but differs in having flowers with stamens attached to the corolla tube (fused petals). Diapensia is a dwarf alpine evergreen with tufts of small, entire leaves crowded on wiry horizontal stems. The 5-merous solitary flowers are held above on short stalks. The creeping plant is found in the Appalachians above treeline only on the highest mountain peaks of Que., NY, and New Eng.; May–Jul.

The only other *Diapensia* species is in the Himalaya Mountains of Asia.

Galax   *Galax aphylla* L.
Diapensia Family   Diapensiaceae

The raceme of small white flowers, 1 to 2 ft. tall, towers over the basal leaves. Although not generally common, Galax is sometimes locally abundant along stream banks in rich mountain woods, especially in NC, SC, GA, and VA; May–Jul.

The thick leathery, roundish leaves, which turn a beautiful bronze in fall, have long been used in floral arrangements. The plant is also sometimes cultivated as a garden ornamental.

Oconee Bells   *Shortia galacifolia* T. & G.
Diapensia Family   Diapensiaceae

This distinctive herbaceous perennial needs little description beyond the photograph. The glossy, spatulate evergreen leaves resemble those of Galax (*above*), thus "galacifolia." Note the flowers, ¾ in. across, with 5 yellow anthers and notched petals. It is an uncommon plant of shaded stream banks in a few scattered mountain counties of NC, SC, and GA; Mar., Apr.

The French botanist Andre Michaux first discovered this plant in w. NC in 1788. Later, in 1839, it was named in honor of botanist Charles Short of KY. It is often cultivated in gardens outside of its original range.

Diapensia
Diapensia lapponica

Galax
Galax aphylla

Oconee Bells
Shortia galacifolia

Carolina Vetch *Vicia caroliniana* Walt.
Pea Family Fabaceae

Vetches are vines that climb by tendrils extending from the ends of pinnately compound leaves. Flowers are located in the leaf axils. Some species are native; others have escaped from cultivation as forage or cover crops.

This native perennial vetch can grow 2–3 ft. long. Note the leaves, which are typical of the genus, and the small (¼ in. long) purple-tipped white flowers. It is common in moist open woods from the Adirondacks and PA southward; Apr.–Jun.

These vetches have purple flowers: Purple Vetch (*V. americana*), May, Jun.; and Bird Vetch (*V. craca*), May–Aug. Hairy Vetch (*V. villosa*) has blue/white bicolored flowers; May–Aug. Yellow Vetch (*V. grandiflora*) has large, dull yellow flowers; Apr.–Jun.

Shooting Star *Dodecatheon meadia* L.
Primrose Family Primulaceae

The unusual swept-back petals, white or pinkish, along with the 5 united stamens, give this plant a distinctive appearance and account for its common name. At the base of the 15- to 25-in.-stalk is a rosette of smooth ovate leaves. It is occasionally found in moist, calcareous woods of the sw. Apps but is more common in midwestern prairies; Apr., May.

American Cowslip (*D. radicatum*) is a smaller plant. Its floral structure is similar but its petals are reddish purple. In the Appalachians, it is apparently restricted to moist, shaded sites along cliffs of the Susquehanna, Schuykill, Ohio, and a few other rivers of PA, WV, and KY; Apr.–Jun.

Star Flower *Trientalis borealis* Raf.
Primrose Family Primulaceae

This low, fragile perennial has a pair of starlike white flowers on a short stalk (4 in. tall) above a whorl of 5–9 sharply pointed leaves. The flowers are unusual in having their parts in 7s. The only member of this family found above treeline, it also occurs in cool northern deciduous forests and high elevation mixed and boreal forests. Its range extends from Que. to VA, WV, and OH; May–Aug.

Carolina Vetch
Vicia caroliniana

Shooting Star
Dodecatheon meadia

Star Flower
Trientalis borealis

Wild Stonecrop   *Sedum ternatum* Michx.
Orpine Family   Crassulaceae

Stonecrops or sedums are low-growing, spreading plants with succulent leaves and small starlike flowers. As their leaves store water, they can and often do endure rocky sites. This, our most common stonecrop, is easily recognized by its small (½ in. across) white flowers and small, smooth leaves. It is seen from NY southward on shaded banks and hillsides, generally on thin rocky soil; Apr.–Jun.

Allegheny Spurge   *Pachysandra procumbens* Michx.
Box Family   Buxaceae

The large (3–4 in. long) leaves of Allegheny Spurge are at the end of the stem, whereas the 2–3 in. flower spikes are attached along the side. It grows in rich calcareous woods along the w. edge of the Apps from KY southward; Mar.–May.

Pachysandra (*P. terminalis*), an Asian plant, is often planted as a ground cover and sometimes escapes. It has smaller, highly variable leaves; flower spikes are at the end of the stems; Apr.–Jun.

Common Dodder   *Cuscuta gronovii* Willd.
Morning-glory Family   Cuscutaceae

Dodders, also known as love vines, are parasitic annuals that lack chlorophyll. They attach themselves to a variety of host plants from which they derive nourishment through haustoria, rootlike connections.

Common Dodder has small (⅛ in. across) white, waxy 5-lobed flowers arranged along yellow or orange stems. It is found in moist, usually disturbed, open habitats throughout our area; Jun.–Nov.

Identification of the other dozen or so species of our area requires careful examination of flowers and fruits. Cherokees poulticed dodder on bruises. Chinese Dodder (*C. chinensis*) is highly valued in Chinese traditional medicine for treating a variety of ailments.

Wild Stonecrop
Sedum ternatum

Allegheny Spurge
Pachysandra procumbens

Common Dodder
Cuscuta gronovii

White Milkweed   *Asclepias variegata* L.
Milkweed Family   Asclepiadaceae

This smooth plant has leaves that are ovate and margins that are either entire or wavy (as seen here). It is a plant of woods and thickets from s. CT and se. NY southward; May–Jul.

Other white-flowered milkweeds are Whorled Milkweed (*A. verticillata*), which has linear leaves in whorls of 3 to 6, Jun.–Sep.; and Tall Milkweed (*A. exaltata*), with leaves pointed toward both ends and sparse, drooping clusters of white flowers, Jun., Jul.

Hairy Mountain-mint   *Pycnanthemum muticum* (Michx.) Pers.
Mint Family   Lamiaceae

Mountain-mints are leafy perennials, typically 1 to 2 ft. tall, with rounded flat-topped clusters of small whitish or grayish flowers. As only a few flowers open at a time, they are not showy. Leaves below the flower clusters are grayish, giving the appearance that they have been dusted with a white powder.

As its name implies, Hairy Mountain-mint is a hairy plant with white to pale lilac flowers. It is found in dry woods and thickets in the mountains from sw. ME southward; Jun.–Aug.

Hoary Mountain-mint (*P. incanum*) is very similar but has prominently toothed leaves that are downy beneath, Jul.–Sep.; Virginia Mountain-mint (*P. virginianum*) has narrower and more pointed leaves that broaden at the base, Jul.–Sep.; Torrey's Mountain-mint (*P. verticillatum*) also has narrow leaves that are slightly toothed and tapering at their base, Jul.–Sep.

Sweet Cicely   *Osmorhiza claytoni* (Michx.) Clarke
Carrot Family   Apiaceae

Plants of the carrot family (here and below) can generally be recognized by their tiny flowers arranged in umbels.

Note the division of the leaves of this 2- to 3-ft. hairy plant into small toothed segments. All parts of the plant, especially the roots, release a distinctive aroma when crushed, accounting for its also being called Aniseroot. It is found in alluvial woods from the Gaspé southward to TN and NC; Apr., May.

John Clayton (?–1773), for whom this species is named, was a pioneer Virginia botanist.

Also called Sweet Cicely, *O. longistylis* is a similar but less hairy plant; May, Jun.

White Milkweed
Asclepias variegata

Hairy Mountain-mint
Pycnanthemum muticum

Sweet Cicely
Osmorhiza claytoni

Water Hemlock (Carotte à Moreau)
Carrot Family

Cicuta maculata L.
Apiaceae

This native biennial, which grows to 6 ft. or more in height, has leaves divided into 3–5 slender, sharp-pointed leaflets. The thick stems are usually spotted with purple. Umbels may be flat-topped, rather than round, as seen here. It is restricted to moist meadows, swamps, and other wet places but is found throughout our area; May–Sep.

All parts of the plant, but especially the roots, are poisonous to humans and livestock (one bite of the root can kill an adult human).

Poison Hemlock
Carrot Family

Conium maculata L.
Apiaceae

Like many other weeds, this tall (to 8 ft.), coarse biennial was introduced from Eurasia. The purple spots on the blue gray stem account for the name *maculata* (spotted). It is common in open, disturbed sites throughout e. U.S.; May, Jun.

As the sap of the plant is poisonous, it was used in ancient Greece to execute criminals, including Socrates.

Queen's Anne's Lace
Carrot Family

Daucus carota L.
Apiaceae

This bristly biennial has flat-topped inflorescences of tiny lacy florets, white except for a single dark purple one near the center. Finely divided fernlike leaves grow from the base of the 2- to 4-ft. stems. Note the old flower cluster that has formed the "bird's-nest" (*upper right*). It is an extremely common roadside weed throughout e. U.S.; May–Oct.

This European plant is apparently the wild ancestor from which the cultivated carrot was developed; thus it is also called Wild Carrot.

Filmy Angelica
Carrot Family

Angelica triquinata Michx.
Apiaceae

The large complex umbels of Filmy Angelica are borne on plants that may reach to a height of 5 ft. Note the stout, reddish stems and the prominent sheaths surrounding the base of the leaf petioles. It often forms large colonies in moist woods or balds at high elevations from PA s. to NC and TN; Jul.–Sep.

Water Hemlock
Cicuta maculata

Poison Hemlock
Conium maculata

Queen Anne's Lace
Daucus carota

Filmy Angelica
Angelica triquinata

Foxglove Beardtongue   *Penstemon digitalis* Nutt.
Figwort Family   Scrophulariaceae

"Beardtongue" refers to the single bearded, sterile stamen that protrudes from each flower of *Penstemon* species. This species, which grows to 3 ft. or more, has sharply pointed opposite leaves. Each of the large (1½ in. long) flowers has 5 conspicuous petals at right angles to the corolla tube. Though less common in the Appalachians than to the west, it is seen in the mountains, principally in open places, from Que. s. to NC and TN; May, Jun.

Slender-flowered Beardtongue (*P. tenuiflorus*) also has slender white flowers but the lower lip of each bends upward toward the upper one; Apr.–Jun.

Culver's-root   *Veronicastrum virginicum* (L.) Farwell
Figwort Family   Scrophulariaceae

Uncommon but conspicuous when present, this tall (to 7 ft.) plant produces several long spikes of tubular flowers, each with two projecting stamens. Note the sharply pointed leaves in whorls of 3–7 per node. It inhabits rich woods from MA and VT s. to NC, SC, and VA; Jun.–Sep.

Also called Culver's-physic because of its medicinal uses, a root tea has been used as a laxative in both Indian and European medicine.

White Wood Aster   *Aster divaricatus* L. var. *divaricatus*
Aster Family   Asteraceae

This common aster has medium-size (2 to 4 in. long) cordate leaves with dentate margins. Flower heads have fewer than 10 rays per head. It grows in woods from s. ME southward. Botanists designate plants of this species that have 10 or more rays and occur at higher elevations of the s. Apps as *A.d.* var. *chlorolepis*; Jul.–Oct.

Other white (or light blue) asters with dentate, cordate leaves include Smooth Heart-leaved Aster (*A. lowrieanus*), which has prominent wings on the petiole, Aug.–Oct.; and Schreber's Aster (*A. schreberi*), which has much larger leaves with basal notches on each side of the leaf base; Jul.–Sep.

Foxglove Beardtongue
Penstemon digitalis

Culver's-root
Veronicastrum virginicum

White Wood Aster
Aster divaricatus

Common Yarrow (Herbe-à-Dindes)   *Achillea millefolium* L.
Aster Family   Asteraceae

The small (½ in. across) white heads of Common Yarrow are arranged into flat-topped clusters. Note the lacy, dissected leaves. This aromatic European weed is scattered across N. Amer. in open waste places; May–Oct.

During ancient times, Common Yarrow was used to treat battle wounds. Both Indians and European settlers used it medicinally to treat ailments such as anorexia, digestive disturbances, colds, and influenza. Recent scientific studies have documented the presence of scores of physiologically active compounds. Cultivated ornamental yarrows are available in several pastel colors.

One may see Alpine Yarrow (*A. borealis*) above treeline of NH peaks. It is a less leafy plant with smaller flower clusters; Jun.–Aug.

Ox-eye Daisy   *Chrysanthemum leucanthemum* L.
Aster Family   Asteraceae

Note the large flower heads to 2 in. across. The prominent depression in the yellow disk separates it from other daisylike plants with white/yellow heads. Growing to 3 ft. tall, this Eurasian weed is found along roadsides and in fields where it can be a serious agricultural pest; Jun.–Aug.

White Snakeroot (Eupatoire Rugueuse)   *Eupatorium rugosum* Houttuyn
Aster Family   Asteraceae

Eupatorium species are generally tall plants with opposite or whorled leaves. Fuzzy flower heads appear in late summer. Other white-flowered eupatoriums are described below.

White Snakeroot is a variable perennial with toothed opposite cordate leaves. It can grow to 5 ft. tall at lower elevations but is a shorter plant at higher elevations. It is found along the edge of woods and in other semi-shaded sites throughout our area; Jul.–Oct.

In the 19th century "milk sickness" was a common and sometimes fatal disease. It was caused by drinking milk from cows that had grazed on this plant. (It is said that President Lincoln's mother died from this poisoning.) American Indians used the roots to treat various ailments including snakebite, thus "Snakeroot."

Common Yarrow
Achillea millefolium

Ox-eye Daisy
Chrysanthemum leucanthemum

White Snakeroot
Eupatorium rugosum

Upland Boneset   *Eupatorium sessilifolium* L.
Aster Family   Asteraceae

Upland Boneset is recognized by its sharply pointed, sessile, opposite leaves. It is found from VT s. to GA, principally in moist, upland woods; Jul.–Oct.

Also called Upland Boneset, *E. altissimum* can be distinguished by its leaves with 3 (vs. 1) main veins; Jul.–Sep. Round-leaved Thoroughwort (*E. rotundifolium*) has round leaves; Jul.–Sep. Boneset (*E. perfoliatum*) has the bases of its leaves joined to surround the stem; Jul.–Oct.

Robin's-plantain   *Erigeron pulchellus* Michx.
Aster Family   Asteraceae

Fleabanes (*Erigeron* species) are downy, often weedy, herbs with alternate, sessile leaves. Flower heads have many thin rays.

This hairy plant, usually less than 2 ft. tall, has showy flower heads; rays may be white or vary from pale lilac to violet. Basal leaves are larger and more rounded than those on the stem above. Spreading by runners, it forms colonies in woods, especially along stream banks from ME southward; Apr.–Jun.

These weedier fleabanes also have white rays, but their leaves are lanceolate: Daisy Fleabane (*E. annuus*), with toothed leaves, May–Oct.; and Common Fleabane (*E. strigosus*), with mostly untoothed leaves, May–Aug.

Wild Quinine   *Parthenium integrifolium* L.
Aster Family   Asteraceae

This smooth perennial (5 ft. tall) has large, rough, prominently toothed leaves. Tiny (¼ in. across) flower heads are arranged in umbels. Also called Eastern Parthenium, it is found in prairies and other open places from ME to GA; May–Jul.

Ozark Parthenium (*P. hispidum*) is a larger plant, with larger heads and hairy stems and leaves; May–Aug.

White Crown-beard   *Verbesina virginica* L.
Aster Family   Asteraceae

Like other members of the genus, this plant is a tall (6–7 ft.), coarse herb with winged stems. Its alternate leaves are broadly lanceolate. It is distinguished from other *Verbesina* species by its flowering heads, which are comprised of gray disk flowers and 3–5 white rays. It occupies woods from VA southward; Aug.–Oct.

The alternative name Frostweed refers to the formation in fall or early winter of "frost flowers" (collarlike ice crystals) around the stems just above the ground. This phenomenon deserves further study.

Upland Boneset
Eupatorium sessilifolium

Robin's-plantain
Erigeron pulchellus

Wild Quinine
Parthenium integrifolium

White Crown-beard
Verbesina virginica

Fraser Magnolia   *Magnolia fraseri* Walter
Magnolia Family   Magnoliaceae

When most people think of "magnolia," they have in mind Southern Magnolia (*M. grandiflora*), the large evergreen tree of the Coastal Plain of se. U.S. In the Appalachians, magnolias are deciduous trees.

Fraser Magnolia has flowers typical of the genus: large and fragrant, with white to cream-colored petals and numerous spirally arranged stamens and pistils in the center. Leaves are auriculate (lobed at base) and green beneath; buds and twigs are smooth. It is common in rich woods, especially along rivers and creeks, from VA and WV s. to GA and SC, at elevations to 4,500 ft.; Mar.–Jun.

Bigleaf Magnolia (*M. macrophylla*) has larger leaves (blades to 3 ft. long), which are auriculate or rounded at the base and silvery underneath; buds and twigs are hairy. It also is found only in the s. Apps; Apr.–Jun.

Umbrella Magnolia   *Magnolia tripetala* (L.) L.
Magnolia Family   Magnoliaceae

The leaves are clustered at the end of twigs in umbrella fashion, and each leaf has a long tapered base. It is found in rich hardwood forests, generally at elevations below 3,000 ft., from s. PA southward; Apr., May.

Cucumber Tree (*M. acuminata*) is also a large forest tree with similar leaves, but they are scattered along the twigs. Flower petals vary from cream to greenish. It is a tree of low elevations found n. to the Adirondacks; Mar.–Jun.

Deciduous magnolias often become large enough for use as timber, but their soft wood limits their use. The outer coverings of their seeds are eaten by wildlife; some are planted as ornamentals.

Large Fothergilla   *Fothergilla major* (Sims) Lodd.
Witch-hazel Family   Hamamelidaceae

Also called Witch-alder, this rare shrub reaches 3 to 4 ft. in height. Note the prominently pinnately veined leaves and showy flower clusters reaching 2 to 3 in. long. It is seen in dry woods and mountain balds of GA, AL, and TN; Apr.–May.

Named in honor of physician-botanist John Fothergill (1712–80) of London, Large Fothergilla is cultivated as an ornamental shrub.

Fraser Magnolia
Magnolia fraseri

Umbrella Magnolia
Magnolia tripetala

Large Fothergilla
Fothergilla major

Silky Camellia
Tea Family

Stewartia malocadendron L.
Theaceae

Also known as Virginia Stewartia, this rare, small (to 15 ft.) deciduous tree has 2- to 4-in. leaves that are softly hairy underneath. The pistil of each flower has 5 fused styles. The spectacular flowers are 2–3 in. across. Found more often on the Coastal Plain, it occurs also at elevations below 800 ft. in the mountains of NC and SC; May, Jun.

Mountain Camellia (*S. ovata*) is a very similar small tree, but it has larger (5–6 in.) leaves and its pistils have separate styles. Its distribution is centered in e. TN, but it also occurs in the mountains of adjacent states; Jun.–Jul.

Sweet Pepperbush
White Alder Family

Clethra acuminata Michx.
Clethraceae

Sweet Pepperbushes are shrubs or small trees with alternate, serrate, deciduous leaves and white flowers in racemes. In this species, found in moist woods from VA and WV southward, the racemes are nonfragrant and pendant; Jul., Aug.

Coast White Alder (*C. alnifolia*) has fragrant, upright racemes. Its range extends from PA to s. New Eng; Jul.–Aug.

Labrador-tea (Lédon du Groenland)
Heath Family

Ledum groenlandicum Oeder
Ericaceae

The heath family includes chiefly shrubs that require acidic soil; many are found in bogs or on mountain balds. Most have urn-shaped flowers with parts in 4s or 5s and simple leaves. (See pink section for additional members.)

This low (to 3 ft.) shrub has thick, alternate entire leaves that are brown and wooly beneath. The small white flowers are in flat clusters. Fruits are slender capsules. A boreal plant common from Greenland across Canada to Alaska, its range extends down the Apps as far s. as PA and OH. Labrador Tea is often associated with alpine meadows but also occurs below treeline in bogs; May–Aug.

The leaves, fragrant when crushed, have been used to make a tea drunk for a great variety of ailments and used externally as a wash for burns and rashes.

Silky Camellia
Stewartia malocadendron

Sweet Pepperbush
Clethra acuminata

Labrador-tea
Ledum groenlandicum

Sand-myrtle *Leiophyllum buxifolium* (Bergius) Elliot
Heath Family Ericaceae

Like Labrador-tea *(above)*, Sand-myrtle is a low-growing, evergreen plant with small white clustered flowers. One notable difference is that leaves are smaller, glossy, and ovate (vs. linear). In the s. Appalachians (TN, NC, SC, and GA), Sand-myrtle is primarily a high-elevation plant (4,500–6,600 ft.) of exposed ridges and summits where it is associated with spruce-fir forests and heath balds. It also occurs, but less commonly, at lower elevations; Apr.– Jun. (sometimes to Oct.).

Sourwood *Oxydendrum arboreum* (L.) DC.
Heath Family Ericaceae

This well-known deciduous tree (to 50 ft.) is quite distinctive. The only species of its genus, no other tree has simple, serrate leaves and small, white, urn-shaped flowers in graceful terminal sprays. In late summer the leaves turn scarlet, while the small whitish fruits are arranged as were the flowers. It is a common tree, especially in oak forests of the mountains of sw. PA and southward at elevations to 2,500 ft.; Jun., Jul.; fruits, Aug.–Nov.

The common name is due to the acidic taste of the leaves, which were used by Native Americans to make a tea for asthma, diarrhea, and indigestion. In KY, European settlers have used it in a similar way and also as a diuretic. Sourwood honey is considered by mountain people to be among the best.

This tree is not to be confused with Sourgum (*Nyssa sylvatica*), a dissimilar lowland tree.

Trailing Arbutus (Fleur de Mai) *Epigaea repens* L.
Heath Family Ericaceae

Also called Mayflower, this evergreen perennial trails on the ground. Note the large smooth, leathery leaves. The generally erect, fragrant trumpetlike flowers, each with 5 flaring petals, may be white or light pink. It is common throughout the Apps, where it is most often found on dry soils in open pine or oak forests; Mar.–Jul.

The plant contains arbutin, a urinary antiseptic; this explains its effective use as a folk and Indian remedy for urinary tract disorders.

Sand-myrtle
Leiophyllum buxifolium

Sourwood
Oxydendrum arboreum

Trailing Arbutus
Epigaea repens

Dog-hobble   *Leucothoe walteri* (Willd.) Melvin *[L. fontanesiana]*
Heath Family   Ericaceae

This evergreen shrub (to 7 ft.) has thick, leathery leaves. Leaves, from 2–6 in. in length, vary in shape but are generally elliptical with toothed margins. Note the arrangement of flower racemes at leaf nodes. It is common in acidic soils, especially along streams, where it often forms thickets so dense that "a dog could hardly hobble through." It is rarely found n. of TN and NC; Apr., May.

Two other *Leucothoe* species, both s. Appalachian deciduous shrubs, share the name Fetter Bush. They are low growing and have flowers similar to those of Dog-hobble: *L. racemosa* grows along lowland stream banks, May, Jun; *L. recurva* is a plant of high elevations, Apr.–Jun.

Bog Rosemary (Andromèda Glauque)   *Andromeda glaucophylla* Link
Heath Family   Ericaceae

A shrub up to 2 ft. high, it is recognized by its narrow, alternate, evergreen leaves, which are greenish blue above and white beneath. White, urn-shaped flowers are similar to those of Dog-hobble *(above)*. Note the rounded, pod-like fruits. Bog Rosemary is primarily a subarctic plant of peatlands of Canada and Greenland, but its range extends at high elevations as far s. as WV; May.

Wintergreen (Petit Thé Bois)   *Gaultheria procumbens* L.
Heath Family   Ericaceae

Another creeping perennial, Wintergreen has glossy, ovate leaves. The few flowers are white and hang like bells. The fruit is a red berry. This aromatic plant occupies sterile woods and clearings throughout our area.

A leaf tea has long been used for the treatment of a variety of common disorders such as rheumatism, fevers, and stomachaches. Methyl salicylate, a powerful essential oil, is apparently the active agent in such uses as well as the cause of its distinct aroma; Jul., Aug.; fruits, Sep.– Nov.

Creeping Snowberry (*G. hispidula*) is another prostrate but more common plant of the n. Apps. It has smaller, ovate leaves, white berries, and only a hint of wintergreen; May, Jun.

The genus was named for physician-botanist Jean-Francois Gaulthier (1708–56) of Que.

Dog-hobble
Leucothoe walteri

Bog Rosemary
Andromeda glaucophylla

Wintergreen
Gaultheria procumbens

Spotted Wintergreen (Herbe à Cle)　　*Chimaphila maculata* (L.) Pursh
Shinleaf Family　　　　　　　　　　　　　　　　　　　Pyrolaceae

Note the whorl of patterned leaves attached to the stem of this small (4–10 in. tall) perennial plant. Although it has the appearance of an herb, its stem is somewhat woody. The nodding flowers are waxy and white to pink. It is common in dry upland forests from s. New Eng. s. to GA; Jun.–Aug.

Pipsissewa (*C. umbellatum*) is distinguished from Spotted Wintergreen by its solid, unpatterned leaves; May–Aug.

The recent discovery in these plants of several compounds with diuretic and antibacterial properties substantiates their long history of use by Native Americans to treat such urinary problems as bladder infections and kidney stones.

Carolina Silverbell　　　　　*Halesia tetraptera* Ellis [*H. carolina*]
Storax Family　　　　　　　　　　　　　　　　　　　Styracaceae

In spring, this large shrub or small tree is easily recognized by its white to pinkish, bell-shaped flowers (1 in. long) arranged along the slender branches. Leaves are 2–7 in. long. At other times of the year, it can be identified by its vertically striped bark or dried fruits with four prominent wings. The only *Halesia* species of the Apps, it is common in rich moist woods below 4,500 ft. from s. OH and WV to GA and AL; Mar.–Jun.

Wild Hydrangea　　　　　　　　　　　*Hydrangea arborescens* L.
Hydrangea Family　　　　　　　　　　　　　　　　　　Saxifragaceae

This large (4 to 10 ft. tall), often straggly shrub has 3-to 6-in.-long, opposite lanceolate leaves. Small white flowers are arranged into a flat-topped or rounded corymb; there are commonly several sterile flowers with much longer petals arranged around the margins of the flower cluster. It is a very common shrub of the s. Apps, where it grows in rich woods usually along stream banks at elevations below 6,000 ft. Its range extends northward to s. NY; Jun., Jul.

Smithsonian Institution ethnographer James Mooney, in his *Myths of the Cherokee,* reports that Cherokees used hydrangea roots to prepare a decoction given to "women who had strange dreams during their menstrual period."

Oak-leaved Hydrangea (*H. quercifolia*), which sometimes escapes from cultivation, has elongated panicles of white flowers and dissected leaves that suggest those of oaks; Jun.–Aug.

Spotted Wintergreen
Chimaphila maculata

Carolina Silverbell
Halesia tetraptera

Wild Hydrangea
Hydrangea arborescens

Ninebark   *Physocarpus opulifolius* (L.) Maxim.
Rose Family   Rosaceae

The rose family is a large temperate one that includes herbs, shrubs, trees, and vines. Flowers have 5 sepals, 5 petals, and numerous stamens and pistils in the center. Others of this family are featured in the pink section.

The common name of this 4- to 10-ft.-tall shrub refers to the bark, which separates into numerous thin layers. Leaves are cordate and somewhat 3-lobed. The typically rosaceous flowers are arranged into compact umbel-like corymbs (see fig. 18). The purple-tinged seed capsules form soon after flowering and are very conspicuous. Ninebark is uncommon, occurring sporadically on stream banks and rocky bluffs throughout the Apps; May–Jul.

Red Chokeberry   *Aronia arbutifolia* (L.) Elliot *[Pyrus arbutifolia]*
Rose Family   Rosaceae

Closely related to pears (*Pyrus*) and apples (*Malus*), chokeberries are also shrubs or small trees; they have simple, alternate leaves and white to pink flowers in a cyme (see fig. 18). The astringent fruits are marginally edible, thus "chokeberry."

Red Chokeberry has hairy twigs and leaves. Note the loose clusters of typical rosaceous flowers. In the fall, it's quite showy with its leaves crimson above and whitish below; fruits are red. It occurs in moist thickets and woods from s. New Eng. southward; Mar.–May.

Black Chokeberry (*A. melanocarpa*) differs by its smooth twigs and leaves, and its black fruits. It occupies somewhat drier sites sporadically throughout the Apps; May–Jul.

Allegheny Serviceberry (Poirier)   *Amelanchier laevis* Wiegand
Rose Family   Rosaceae

In isolated mountain communities of earlier times, funeral services for people who had died during the winter were postponed until early spring when these trees bloom; thus "serviceberry," often pronounced "sarvisberry." The smooth, serrated leaves are only partially unfolded when the flowers with their 5 long petals first bloom. The ripe fruits are dark purple, juicy, and delicious. Plants occur in a variety of habitats, especially in deciduous forests of the s. Apps where it ranges from lowlands to 6,000 ft. It also extends northward to Que. at decreasing elevations; Mar.–Jun.; fruits, May–Aug.

Downy Serviceberry (*A. arborea*) is also a common and widespread species. In contrast to *A. laevis*, it has hairy leaves that are fully open during flowering and fruits that are dark red and dry; Apr.–Jun.; fruits, Jun.–Aug.

Ninebark
Physocarpus opulifolius

Red Chokeberry
Aronia arbutifolia

Allegheny Serviceberry
Amelanchier laevis

Blackberry (Ronce)   *Rubus* species
Rose Family   Rosaceae

Identification of the numerous (200 or more?) species of *Rubus*, collectively "brambles," is difficult. One useful division is into blackberries, dewberries, and raspberries.

Blackberries have erect or arching canes (stems) 3–8 ft. long. Their compound leaves typically have 5 leaflets on first-year canes and 3 on second-year canes. Flowers, about ½ in. across, are followed by red fruits that turn black as they ripen.

Blackberries, along with other brambles, are often abundant in old fields (i.e., open areas undergoing ecological succession). Various species are found throughout the Apps; May–Jul.; fruits, Jul.–Sep.

Prickles, common on canes of low-elevation plants, are absent on those from higher elevations.

American Mountain-ash (Cormier)   *Sorbus americana* Marshall
                                        [Pyrus americana]
Rose Family   Rosaceae

Closely related to pears *(Pyrus)* and apples *(Malus)*, but not to ashes *(Fraxinus)*, mountain-ashes are shrubs or small trees. This species, which may become 30 ft. tall, has compound leaves with long (2–4 in.) lanceolate leaflets. Small, white flowers occur in dense, flat-topped cymes that produce large, bright orange (or red) fruit clusters. Fruits are not produced every year, but rather at 3-year intervals. This conspicuous tree occurs in the spruce-fir forests of the s. Apps and northward to Que. at increasingly lower elevations; May–Jul.; fruits, Aug.–Oct.

Common Hawthorn (Pommettes)   *Crataegus flabellata* (Bosc.)
                                        K. Koch. *[C. macrosperma]*
Rose Family   Rosaceae

Hawthorns are a group of gnarly shrubs or small trees that are widespread in open or exposed sites throughout the Appalachians. Unlike other rosaceous woody plants, they have long (1–5 in.), unbranched spines attached to the twigs of older branches. Leaves are simple, serrate, and either lobed or unlobed. As it takes a specialist to identify to species (taxonomists can't even agree how many there are!), no attempt is made here to distinguish *C. flabellata* from at least 100 other mountain species.

Common Hawthorn, with its lobed leaves, white-petaled flowers followed by small red applelike haws (fruits), is a typical species. Found throughout our area, it is probably the most common hawthorn of the s. Apps; May–Jul.; fruits, Jul.–Nov.

Blackberry
Rubus species

American Mountain-ash
Sorbus americana

Common Hawthorn
Crataegus flabellata

Black Cherry (Cerisier Tardif)   *Prunus serotina* Ehrh.
Rose Family   Rosaceae

This large (to 75 ft.) tree has glossy, simple leaves with finely toothed margins. Flowers, ⅜ in. wide, are in elongate racemes. Fruits, which turn from red to black, are bitter but eaten by wildlife. Found in deciduous forests throughout the Apps and common along roadsides; Apr., May; fruits, Jun.–Aug.

These are also common but smaller trees: Chokeberry (*P. virginiana*) has similar flower clusters but shorter, more rounded leaves, Apr.–Jul.; fruits, Jul.–Oct; Fire Cherry (*P. pensylvanica*) has flowers (and fruits) in smaller, flattened clusters and narrower leaves, Mar.–Jul.; fruits, Jul.–Sep.

Black Locust   *Robinia pseudo-acacia* L.
Pea Family   Fabaceae

The showy racemes (4 to 8 in. long) of fragrant, white, pealike flowers appear just before the compound leaves, with their 3–10 pairs of ovate leaflets, unfold. Black Locust is a large, often asymmetrical tree with short, paired thorns on the branches (but not the trunk). Common in thickets and woods, it was originally native to the mountains n. to PA but, due to planting, is now naturalized as far n. as Que.; May, Jun.

Yellowwood (*Cladrastis kentuckea*) is a tree of the s. Apps with white racemes similar to those of Black Locust. It is thornless and its leaves have fewer, wider leaflets.

New Jersey Tea   *Ceanothus americanus* L.
Buckthorn Family   Rhamnaceae

This low-growing shrub (to 4 ft.) bears complex clusters of tiny, 5-petaled flowers in the axils of its 2- to 4-in.-long, serrate, opposite leaves. It is a very common wildflower of open, often dry, woods and fields n. to s. Que.; May–Sep.

American Mistletoe   *Phoradendron serotinum* (Raf.)
   M. C. Johnston [*P. flavescens*]
Christmas-mistletoe Family   Viscaceae

This semiparasitic shrub grows attached to the branches of various deciduous hardwoods, such as maple, oak, hickory, and black gum. The evergreen leaves of mistletoe are small (2–3 in. long) and oblong. Small white flowers produced by the female plants in late fall are followed by white winter berries. This, the only Appalachian mistletoe, is quite common in the south, becoming less so northward. Its northernmost limit in OH and WV is apparently determined by its inability to survive harsh winters. For the same reason, it is limited to low and middle elevations of the mountains; Oct., Nov.; fruits, Nov.–Jan.

Black Cherry
Prunus serotina

Black Locust
Robinia pseudo-acacia

New Jersey Tea
Ceanothus americanus

American Mistletoe
Phoradendron serotinum

Flowering Dogwood   *Cornus florida* L.
Dogwood Family   Cornaceae

This, our only bracted dogwood tree, is the well-known and showy native dogwood, which is also widely planted as an ornamental. The horizontal branches bear alternate, simple leaves and inflorescences that consist of a cluster of inconspicuous greenish yellow flowers surrounded by 4 large white (or pink) notched bracts. Fruits are bright red, elongated berries. Flowering Dogwood is an understory tree (to 35 ft.) in dry, acidic soils of oak-pine forests from s. New Eng. southward, but it is more common in the s. Apps; Mar.–Jun.; fruits, Aug.–Nov.

The tree was regarded by Indians, and also by some early African and European settlers, as a virtual "drugstore." Twigs were chewed as a substitute for brushing one's teeth; root-bark was used to treat malaria; and the bitter berries for stomach upset. The bark can be used to produce a scarlet dye, as well as a black ink. The hard, close-grained wood is useful in the textile industry for shuttles and bobbins. Unfortunately, Flowering Dogwoods throughout the Apps are infected or threatened by anthracnose, a serious fungal disease.

Southern Swamp-dogwood   *Cornus stricta* Lam. [*C. foemina*]
Dogwood Family   Cornaceae

Bractless dogwoods are shrubs or small trees with simple leaves and clusters of inconspicuous green or white flowers that lack showy bracts. This dogwood has smooth, opposite leaves, maroon twigs, and small, white flowers in round-topped clusters. Fruits are blue. It occupies wet sites of the s. Appalachians n. to VA and WV; May, Jun.; fruits, Aug.–Oct.

Northern Swamp-dogwood (*C. racemosa*) can be distinguished by its brownish twigs. Other bractless dogwoods include Red-osier Dogwood (*C. sericea*), with rough leaves, reddish twigs, and white fruits, May–Aug.; fruits, Aug.–Oct.; and Alternate-leaved Dogwood (*C. alternifolia*), our only dogwood with alternate leaves; fruits are blue; May, Jun.; fruits, Aug.–Oct.

Fringe Tree   *Chionanthus virginicus* L.
Olive Family   Oleaceae

The Fringe Tree is a showy shrub or small tree; its clusters of flowers, each with 5- to 6-in.-long, slender petals, account for the alternative name, Old-man's-beard. Fruits resemble ripe olives. Leaves are simple, elliptical to obovate, and 3–8 in. long. It is a tree of moist woods and bluffs of se. U.S. More common on the Coastal Plain, it occurs sporadically in the Apps from s. PA southward. It is sometimes cultivated n. of its natural range; May, Jun.

Flowering Dogwood
Cornus florida

Southern Swamp-dogwood
Cornus stricta

Fringe Tree
Chionanthus virginicus

Witch-hobble (Bois d'Orginial)   *Viburnum alnifolium* Marshall
                                                          [*V. lantanoides*]
Honeysuckle Family   Caprifoliaceae

Viburnums are shrubs with large, opposite, deciduous leaves that are usually toothed or lobed. Flowers are small and white, have 5 petals, and are arranged into compound cymes. The fruit is a 1-seeded, black or red drupe.

Also called Hobble-bush, the 10-ft.-tall shrub has round, scalloped leaves. Flowers include larger peripheral ones, as well as the smaller ones in the center of the cluster. Berries turn from red to black. Found throughout our mountains, it is a shrub of cool woods; May, Jun.; fruits, Aug.–Sep.

Maple-leaf Viburnum   *Viburnum acerifolium* L.
Honeysuckle Family   Caprifoliaceae

The leaves of this shrub resemble those of maple *(Acer)* in both shape and their opposite arrangement on the stem. The white to cream flower clusters, 1–3 in. across, are followed in fall by fleshy purplish black, berrylike fruits. At this time the leaves turn delightful shades of pink, lilac, or magenta. This plant is common throughout the Apps; May–Jul.; fruits, Aug.–Oct.

Other viburnums also lack the showy peripheral flowers of Witch-hobble *(above)*. *Viburnum dentatum* (Jun.–Aug.) and *V. rafinesquianum*, both called Arrow-wood, have dentate leaf margins; leaves of the latter are sessile; May, Jun. Black Haw *(V. prunifolium)* has serrate, ovate leaves; Apr., May.

Common Elderberry (Sureau du Canada)   *Sambucus canadensis* L.
Honeysuckle Family   Caprifoliaceae

This common shrub grows to a height of 6–8 ft. Its showy, flat-topped flower clusters include tiny, white flowers. Large leaves are divided into 5–11 coarse-toothed leaflets. Elderberry occupies a variety of moist sites throughout the Apps, but it is especially common in deciduous forests at lower elevations; Jun., Jul.; fruits, Aug.–Oct.

The flowers can be used to prepare fritters; the purple fruits, to make jelly or wine.

Red Elderberry *(S. pubens)* is a similar shrub that has rounded flower clusters and red fruits; it occupies high-elevation woods; May, Jun.; fruits, Aug.–Oct.

Witch-hobble
Viburnum alnifolium

Maple-leaf Viburnum
Viburnum acerifolium

Common Elderberry
Sambucus canadensis

Poison Ivy (Here à la Puce) *Toxicodendron radicans*
(L.) Kuntze *[Rhus radicans]*
Cashew Family Anacardiaceae

So variable is this widespread vine that some botanists have divided it into more than 30 species. It may trail on the ground or climb high into trees. Note the glossy leaves (bright red in the fall) with their 3 leaflets and small, white flowers; the white berries are eaten by birds, apparently with no ill effects. It grows best in moist, semishaded sites, but it can grow in various low-elevation habitats throughout our mountains; May, Jun.; fruits, Jul.–Nov.

At least 50 percent of people will develop a painful dermatitis if exposed to the active ingredient, toxicodendrol, found in all parts of the plant.

Poison Oak (*T. pubescens*) also has trifoliate leaves, but the leaflets are lobed; also, it is more often an upright shrub of dry places; May. Poison Sumac (*T. vernix*) is a shrub of swamps; it has 7 or more leaflets per leaf; May, Jun. These two species, too, may cause a dermatitis.

American Holly *Ilex opaca* Ait.
Holly Family Aquifoliaceae

Hollies are deciduous or evergreen shrubs or small trees. They are dioecious; the staminate (male) and pistillate (female) flowers are produced on separate plants. Fruits of both species pictured here are shown in the red section.

American Holly is our only evergreen species of *Ilex*. The bark of the medium-size tree (to 40 ft.) is smooth and dark gray. Leaves are dark green, leathery, and have prickly margins. Primarily a tree of southeastern lowland forests, it is also frequent in low to middle elevations as far north as WV; Apr.–Jun.; fruits, Jun.–winter.

This is the holly most often used in Christmas decorations.

Winterberry (Houx Verticillé) *Ilex verticillata* (L.) A. Gray
Holly Family Aquifoliaceae

The leaves (2–4 in. long) of this deciduous shrub or small tree (to 25 ft.) are variable, often appearing as seen here, but sometimes with dentate margins. Small white flowers (and red berries) are on short ($\frac{1}{2}$ in. or less) stalks. Winterberry is found on moist to wet sites throughout the Apps except for Quebec's highest peaks; Jun.; fruits, Oct.–Jan.

Mountain Holly (*I. montana*) is also deciduous, but it has larger leaves (4–6 in. long) and its flowers and berries occur on longer stalks; May, Jun.; fruits, Oct.–Jan.

Poison Ivy
Toxicodendron radicans

American Holly
Ilex opaca

Winterberry
Ilex verticillata

Yellow

Golden Club   *Orontium aquaticum* L.
Arum Family   Araceae

Arums are characterized by a thick stalk bearing tiny flowers (the spadix), surrounded by a tubular, sheathing leaf (the spathe). Although most species are tropical, several have ranges that extend northward into the Apps. Other arums are featured in the green/brown section.

The ft.-long spadices of Golden Club bear tiny, yellow, bisexual flowers (spathes are shed earlier). Another name, Neverwet, is due to its waxy, water-repellent leaves that float or extend above the water. Uncommon, but conspicuous when present, this emergent perennial is found in swamps and shallow ponds from MA southward; Apr.–Jun.

Yellow Trillium   *Trillium luteum* (Muhl.) Harbison
Lily Family   Liliaceae

The lily family is a large one that is well represented in temperate regions. Most lilies are perennials that survive the winter as dormant bulbs. Flowers typically have 3 sepals, 3 petals, 6 stamens, and a single pistil. Other members of the family are shown in the white and the red sections.

Yellow Trillium is a variable species, but it is easily recognized by its green to yellow lanceolate petals and strongly mottled leaves. It is common on basic soils in woods at low elevations of the Smoky Mountains, but it is seldom found n. of TN and NC. The flowers have a very pleasant lemon scent; Apr., May.

Pale Yellow Trillium (*T. discolor*), found only in the drainage area of the upper Savannah R., has cream to pale yellow petals rounded on the ends; May.

Trout-lily (Ail Soux)   *Erythronium americanum* Ker Gawler
Lily Family   Liliaceae

The mottled leaves (often more prominently marked than here), together with the 6 yellow tepals (sepals and petals collectively), help to identify this 4- to 10-in.-tall plant. Anther color varies from yellow to bright cinnamon. One of the first wildflowers to bloom, it is quite showy in large colonies, often on basic soils, from s. Canada southward. It grows in low-elevation deciduous forests, as well as high-elevation boreal ones; Mar.–Jun.

Both roots and leaves were used by Native Americans in medicinal preparations for various ailments. Modern research indicates that water extracts of Trout-lily are antibacterial.

White Trout-lily (*E. albidum*) is similar, but it has white flowers and unmottled leaves; Mar.–May.

Golden Club
Orontium aquaticum

Yellow Trillium
Trillium luteum

Trout-lily
Erythronium americanum

Indian Cucumber-root (Jarnotte)   *Medeola virginiana* L.
Lily Family   Liliaceae

This (1–2 ft.) plant, with its 2 whorls of leaves and small dangling flowers, is quite distinctive. Fruits are purple or black berries. A shade-tolerant perennial, it is common in well-drained forests throughout the Apps; Apr.–Jun.; fruits, Jul.–Sep.

Its cucumber-flavored roots are delicious (but its scarcity should prevent one from eating them, except in an emergency). A root tea has been used as a diuretic; a tea made from leaves and berries, to prevent convulsions in babies.

Clinton's-lily   *Clintonia borealis* (Ait.) Raf.
Lily Family   Liliaceae

The 2–3 broad, glossy leaves occur at the base of the 12- to 14-in.-tall flower stalk that bears yellow flowers. It's also called Bluebead Lily because of its bright blue berries. This shade-tolerant plant is found in both deciduous and spruce-fir forests, often at high elevations from s. Que. southward; May–Aug; fruits, Jul.–Oct.

Speckled Wood-lily (*C. umbellulata*) has an umbel of smaller, spotted greenish white flowers that form black berries; May, Jun.; fruits, Jul.

DeWitt Clinton (1769–1828) was a governor of New York.

Large-flowered Bellwort   *Uvularia grandiflora* J. E. Smith
Lily Family   Liliaceae

Note the relatively large (¾–1 in. long) flowers of this 1- to 1½-ft.-plant; also note the perfoliate (pierced by stem) leaves, which are hairy underneath. It occurs in rich, often limestone-derived soils in the mountains from s. Que. southward; Apr.–Jun.

The following are similar Appalachian wildflowers that are also called bellwort or wild-oats but that have smaller yellow flowers: *U. perfoliata*, like above but with leaves that are smooth underneath, May, Jun.; *U. sessilifolia*, leaves sessile, not glossy, Apr., May; and *U. puberula*, leaves sessile but glossy, May.

American Indians used root teas from *Uvularia* species for a great variety of ailments, but little or no modern research has been done to test their efficacy.

Indian Cucumber-root
Medeola virginiana

Clinton's-lily
Clintonia borealis

Large-flowered Bellwort
Uvularia grandiflora

Yellow Mandarin *Disporum lanuginosum* (Michx.) Nicholson
Lily Family Liliaceae

Yellow Mandarin is a downy 1- to 2-ft.-tall plant with alternate leaves and unspotted yellow green flowers; its fruit is a red berry. It is found in rich mountain woods from the Adirondacks and New Eng. southward; May, Jun.; fruits, Jun.–Sep.

The less common Spotted Mandarin (*D. maculatum*) has purple-spotted, white or yellowish sepals and petals. It is found from e. OH southward; Apr., May; fruits, May–Sep.

Yellow Star-grass *Hypoxis hirsuta* (L.) Coville
Lily Family Liliaceae

The 6- to 12-in. grasslike leaves of this hairy perennial are longer than the stems that bear the small cluster of ½-in.-wide starlike flowers. Each flower has 3 yellow petal-like sepals, 3 yellow petals, and 6 stamens. It is fairly frequent in open woods, along stream banks, and in meadows from the Adirondacks and New Eng. southward; Apr.–Sep.

Yellow Iris *Iris pseudacorus* L.
Iris Family Iridaceae

This 3-ft.-tall iris, of European origin, often escapes cultivation and forms large colonies in marshes and streams throughout our area. It is also called Yellow Flag and Water Flag; May, Jun.

From this plant can be obtained both yellow (flowers) and brown or black (root) dyes. It was widely used in Europe for various medicinal uses; pharmacological evidence indicates that it has some value as an anti-inflammatory agent, but most of its uses have not been confirmed.

Large Yellow Lady's-slipper (Cypripède Soulier) *Cypripedium calceolus*
L. var. *pubescens* (Willd.) Correll
Orchid Family Orchidaceae

Showiest of our orchids are the lady's-slippers or moccasin flowers, *Cypripedium* species. The conspicuous slipperlike pouch formed by the lower petal is a trap for capturing bees, which are released only after being coated with pollen. American Indians used the roots of these plants, as did 19th-century physicians, for many types of nervous ailments such as hysteria, insomnia, and premenstrual syndrome.

Large Yellow Lady's-slipper is a hairy plant that grows to 2½ ft. It grows in moist deciduous woods from s. New Eng. southward; Apr.–Jun.

Small Yellow Lady's-slipper (*C. calceolus* var. *parviflorum*) is a smaller plant (to 8 in.) with twisted purple petals; it is more common in bogs of the n. Apps; Apr.–Jun.

Yellow Mandarin
Disporum lanuginosum

Yellow Star-grass
Hypoxis hirsuta

Yellow Iris
Iris pseudacorus

Large Yellow Lady's-slipper
Cypripedium calceolus

Tulip-tree   *Liriodendron tulipifera* L.
Magnolia Family   Magnoliaceae

This tree has leaves whose sharp-tipped lobes resemble those of some maples. Note also the flowers, each with 3 reflexed sepals and 6 petals marked with orange. This tree reaches its largest size in cove hardwood forests of the s. Apps, but it also invades less desirable sites, where it often forms extensive successional stands following fires or logging. It is among the most common mountain trees from New Eng. southward; Apr.–Jun.

The bark and roots were used medicinally by Native Americans for a variety of ailments including malaria, rheumatism, and arthritis; it has also been used as an aphrodisiac. The soft, but durable, yellowish wood of this tree is used for furniture, millwork, and pulp. It is known in the lumber trade as "yellow-poplar," but it should not be confused with true poplars (*Populus* species). Tulip-tree is the state tree of TN.

Spatterdock (Lis d'Eau Jaune)   *Nuphar microphylla* (Pers.) Fern.
Water-lily Family   Nymphaeaceae

Spatterdock has ovate leaves, which may be submerged or above water, with rounded notches and small (1–2 in. across) yellow flowers. It is found in ponds and lakes throughout most of our area.

The roots of Spatterdock were a source of medicine for American Indians, who used a tea to treat many ailments. The roots contain a variety of biologically active alkaloids and steroids; May–Oct.

Yellow Lotus (*Nelumbo lutea*) has much larger flowers (6 to 8 in. wide) and a large pistil resembling a saltshaker; Jun.–Aug.

Celandine Poppy   *Stylophorum diphyllum* (Michx.) Nutt.
Poppy Family   Papaveraceae

Below each cluster of yellow-petaled flowers (1½–2 in. across) is a single pair of large, deeply dissected leaves with rounded lobes. Other leaves are basal. Seed pods are ovoid and hairy. Also called Wood Poppy, it is found in moist woods from w. PA s. to TN; Mar.–May.

The alien Celandine (*Chelidonium majus*), or Herbe aux Verrues, of the same family presents a similar appearance. However, its leaves are attached singly to the stem; its flowers (¾ in. across) are smaller; and its fruit is a smooth, slender seed pod; Apr.–Sep.

Tulip-tree
Liriodendron tulipifera

Spatterdock
Nuphar microphylla

Celandine Poppy
Stylophorum diphyllum

Marsh-marigold (Souci d'Eau)   *Caltha palustris* L.
Buttercup Family   Ranunculaceae

This aquatic perennial grows to 2 ft. and has thick, hollow stems. Leaves are glossy and round or reniform (kidney-shaped). These features, together with its golden flowers (5 to 9 sepals each), make it a strikingly handsome plant. Introduced from Europe, it is now common in marshes and swamps throughout our area; Apr.–Jun.

A root tea has been used medicinally by both Indians and Europeans, but poisoning can result from eating the raw plant.

Hooked Buttercup (Renoncule)   *Ranunculus recurvatus* Poiret.
Buttercup Family   Ranunculaceae

About 20 species of buttercups (*Ranunculus* species) are found in Appalachia. Herbs with deeply cut, alternate leaves, they often inhabit moist to wet, open waste places. Flowers have varying numbers of yellow sepals and petals and numerous bushy stamens. Buttercups are generally poisonous if ingested; the sap may also be a skin irritant.

This common 1- to 2-ft.-tall buttercup is a hairy plant with inconspicuous flowers. The green, recurved sepals are longer than the small yellow petals. "Hooked" refers to the beaks on the achenes (small, single-seeded fruits); May–Jul.

Bulbous Buttercup   *Ranunculus bulbosus* L.
Buttercup Family   Ranunculaceae

Like Hooked Buttercup (*above*), this species is also hairy and has reflexed sepals. It is a smaller (1–1½ ft. tall) plant, however, with larger, showier flowers. Its name is derived from the bulblike swelling of its roots. An alien species, it is found in moist places throughout our area; Apr.–Aug.

Also an alien, Common or Tall Buttercup (*R. acris*) is a more common and taller (2–3 ft.) plant. Its sepals are spreading (vs. reflexed), and its roots are not bulbous; May–Sep.

Marsh-marigold
Caltha palustris

Hooked Buttercup
Ranunculus recurvatus

Bulbous Buttercup
Ranunculus bulbosus

Witch-hazel (Cafe du Diable)   *Hamamelis virginiana* L.
Witch-hazel Family   Hamamelidaceae

This deciduous shrub or small tree (to 20 ft. tall) has rounded leaves with wavy or toothed margins. Appearing in late autumn, the flowers open as the previous year's fruits ripen; each flower includes 4 long, twisted petals as seen here. Note also that at flowering time most of the leaves have already been shed. Witch-hazel grows in dry to moist woods, often on slopes, throughout the Apps from s. Que. southward; Sep.–Dec.

A distillate of Witch-hazel is used externally as an astringent. A traditional way of locating, or "witching" for, underground water has been to use a divining rod made from a forked branch of this shrub.

Smooth Yellow Violet   *Viola eriocarpa* Schw.
Violet Family   Violaceae

For an introduction to violets, including the distinction between "stemmed" vs. "stemless" violets, see the blue/purple section.

This stemmed yellow violet has broadly heart-shaped leaves; both the stems and leaves are smooth. It is found in meadows and low-elevation forests from Que. southward; Apr.–Jun.

Downy Yellow Violet (*V. pubescens*) is also a stemmed yellow violet but has narrower leaves; its stems and leaves are covered with fine hairs; Apr., May.

Our only stemless yellow violet is Round-leaved Violet (*V. rotundifolia*); Apr., May.

Halberd-leaved Yellow Violet   *Viola hastata* Michx.
Violet Family   Violaceae

The mottled, distinctively shaped leaves make this stemmed violet easy to recognize. It grows in deciduous forests from PA southward; Apr., May.

A halberd is an axlike weapon used during the 15th and 16th centuries.

Three-part-leaved Violet (*V. tripartita*) is a yellow, stemmed violet identified by its variable, but generally 3-lobed, leaves; Apr., May.

Witch-hazel
Hamamelis virginiana

Smooth Yellow Violet
Viola eriocarpa

Halberd-leaved Yellow Violet
Viola hastata

Roan Mountain Avens   *Geum radiatum* Michx.
Rose Family   Rosaceae

Avens (*Geum* species) are small plants with large, simple or compound leaves; their cup-shaped flowers have 5 petals that are usually yellow, but in some species are pink, white, or purple. The roots and leaves of various Asian and North American avens have been used to prepare teas with diuretic and astringent properties.

In Roan Mountain Avens, the lobed simple leaves are 5–6 in. across; the long-stemmed flowers, 1½ in. wide, resemble those of buttercups. It is an endangered species that is restricted to a few exposed, rocky, high-elevation sites in NC and TN such as Roan Mountain; Jun., Jul.

White Mountain Avens   *Geum peckii* Pursh
Rose Family   Rosaceae

Closely related to Roan Mountain Avens (*above*), White Mountain Avens is restricted to a few alpine mountain areas of Nova Scotia and the White Mountains of NH. There, its large flowers (1–2 in. across) make it the showiest of plants above treeline. Note also the large, fan-shaped leaves with margins; Jun.–Sep.

Purple Avens (*G. rivale*) has smaller, nodding flowers, typically in clusters of 3; the color of its sepals and petals varies from purple to yellow. Leaves are divided irregularly into several segments. It is a bog plant of lower elevations from Que. s. to WV. The name "Chocolate-root" refers to the use of Purple Avens in preparing a drink similar to hot chocolate; May.–Aug.

Large-leaved Avens   *Geum macrophyllum* Willd.
Rose Family   Rosaceae

In contrast to the other avens pictured above, this one is a hairy plant and has even larger leaves (up to a foot across). Its smaller flowers, each consisting of 5 narrow and nonoverlapping petals, are less cup-shaped. It is a plant of Que. and n. New Eng., where it occurs in moist woods and thickets; May–Jul.

Rough Avens (*G. virginianum*) has similar, but pale yellow, flowers and leaves divided into 3 main leaflets. It is found from the Adirondacks s. to TN and SC; Jun., Jul.

Roan Mountain Avens
Geum radiatum

White Mountain Avens
Geum peckii

Large-leaved Avens
Geum macrophyllum

Dwarf Cinquefoil   *Potentilla canadensis* L.
Rose Family   Rosaceae

"Cinquefoil" refers to the 5 leaflets characteristic of many *Potentilla* species; in many cases, however, leaves have 7 leaflets. Flowers are usually yellow and resemble those of avens *(above)*.

Dwarf Cinquefoil has 5 wedge-shaped leaflets, each with prominent teeth at its apex. Like most cinquefoils, it is a low plant that spreads by runners. Fairly common throughout most of our area, it grows in open woods and fields; Apr.–Jun.

Common Cinquefoil (*P. simplex*) is a larger plant that can be distinguished by its leaflets, which taper toward their tips. It is also common throughout the Apps; Apr.–Jun.

Silver-weed (Argentine)   *Potentilla anserina* L. *[Argentina anserina]*
Rose Family   Rosaceae

This low-growing, spreading potentilla has leaves and flowers on separate stalks. Its most distinctive feature is its leaves, which are silvery below and divided into many leaflets, each strikingly toothed. A circumboreal species, Silver-weed has a range that extends s. to the Adirondacks. It is more common on sandy coasts but also occurs in low-elevation mountain sites; May–Sep.

The large roots of mature plants may be cooked and eaten like turnips.

Sulfur Cinquefoil   *Potentilla recta* L. *[P. sulphurea]*
Rose Family   Rosaceae

"Sulfur" reflects the pale yellow of the petals. This is a taller (to 2 ft.), coarser plant than other cinquefoils featured here. Leaves are divided into 5–7 toothed, relatively narrow leaflets. Also called Rough-fruited Cinquefoil, this alien plant grows throughout our area, especially along roadsides and other disturbed sites; Jun.–Aug.

Rough Cinquefoil (*P. norvegica*) is also an upright native plant; it differs primarily in having its leaves divided into 3 leaflets. It is also widespread and common; Jun.–Oct.

Dwarf Cinquefoil
Potentilla canadensis

Silver-weed
Potentilla anserina

Sulfur Cinquefoil
Potentilla recta

Golden St. John's-wort (Milleper Tuis) *Hypericum frondosum* Michx.
St. John's-wort Family Clusiaceae

Among St. John's-worts (*Hypericum* species) are herbs and shrubs bearing simple, opposite leaves with entire margins; leaves are often dotted with minute glands. Flowers have 5 (or 4) yellow petals and numerous stamens. The common name was given because they flower in midsummer at the time of the birthday of St. John the Baptist.

This shrub (to 3 ft.) is distinguished by its large (1 to 1½ in. across) bushy flowers. It is often locally abundant on thin, calcareous soils along the w. edge of the Apps of KY and TN; elsewhere, it is cultivated as an ornamental; Jun., Jul.

Other *Hypericum* shrubs include Bushy St. John's-wort (*H. densiflorum*), which has clusters of ½-in. flowers, Jul.–Sep.; and Shrubby St. John's-wort (*H. spathulatum*), with flowers almost as large as those of *H. frondosum* but with narrower leaves, May–Jul.

Common St. John's-wort (Pertuisane) *Hypericum perforatum* L.
St. John's-wort Family Clusiaceae

This European plant, our only alien St. John's-wort, is a highly branched, 1- to 2-ft.-tall herb that bears numerous flowers to 1 in. across. The petals have black dots along their margins. It occurs at lower elevations throughout the Apps, especially along roadsides and other disturbed sites; Jun.–Sep.

There is a renewed medicinal interest in this plant. Recent studies indicate that its extracts are generally superior to synthetic drugs as an antidepressant.

Spotted St. John's-wort (*H. punctatum*) is similar, but it has prominent black dots on its leaves as well as on its flower petals, and its flowers are smaller; Jun.–Sep.

Mountain St. John's-wort *Hypericum mitchellianum* Rydb.
St. John's-wort Family Clusiaceae

This low-growing (1–2 ft.), largely unbranched plant has conspicuous flowers with amber petals and long stamens. It grows in high-elevation meadows of VA, TN and NC; Jun.–Aug.

Hypericum buckleyi is also a s. Appalachian endemic of high elevations, including balds and rocky crevices. It can be distinguished from Mountain St. John's-wort by its rounded leaves and lemon yellow flowers; Jun.–Aug.

Golden St. John's-wort
Hypericum frondosum

Common St. John's-wort
Hypericum perforatum

Mountain St. John's-wort
Hypericum mitchellianum

Partridge-pea   *Chamaecrista fasciculata* (Michx.)
                                 Greene *[Cassia fasciculata]*
Caesalpinia Family   Caesalpiniaceae

Chamaecrista species are yellow-flowered, erect, annual herbs with compound leaves. Partridge-pea (1 to 3 ft. tall) has 5 to 6 pairs of bristle-tipped leaflets per leaf, and 1-in.-wide flowers with 5 uneven petals and dark, curved stamens. It is common in open fields and waste places n. to MA. (Note how well the segmented abdomen of the praying mantis blends in with the leaves); Jul.–Sep.

Also common, Wild Sensitive-plant (*C. nictitans*) is a much smaller plant that produces smaller flowers and leaves that fold when touched; Jul.–Sep.

Common Evening-primrose   *Oenothera biennis* L.
Evening-primrose Family   Onagraceae

Evening-primroses (not related to primroses, Primulaceae) are herbs with a distinctive flower: 4 reflexed sepals and 4 petals attached at the end of a long calyx tube; the stigma has 4 branches that form a cross. The common name is derived from the several species that open toward evening (others, called sundrops, open earlier in the day).

This tall (to 5 ft.), stately plant bears numerous simple, pointed leaves along its hairy, often reddish stem. The flowers, typical of the genus, are generally 1 to 1½ in. across. A transcontinental species, it is common in disturbed, open sites from s. Que. southward; Jun.–Oct.

There is currently considerable interest in this plant as the source of a seed-oil that appears to be useful in the treatment of such diverse disorders as alcoholism, asthma, and premenstrual syndrome. Indians used the roots medicinally.

Shale Barren Evening-primrose (*O. argillicola*), also a tall plant with large yellow flowers, occurs primarily on shale barrens; Jul., Aug.

Southern Sundrops   *Oenothera fruticosa* L. *[O. linearis]*
Evening-primrose Family   Onagraceae

This variable, often hairy plant, which reaches 2 or 3 ft. in height, has lance-shaped leaves; flowers are 1 in. or more across. Stamens are often orange. It grows in dry soils of open fields and woods throughout our area. As it is cultivated, it probably occurs also as a garden escape; May–Aug.

Northern Sundrops (*O. tetragona*) is similar, but generally has smooth leaves and stems and usually occurs at higher elevations; May–Aug.

Partridge-pea
Chamaecrista fasciculata

Common Evening-primrose
Oenothera biennis

Southern Sundrops
Oenothera fruticosa

Garden Loosestrife   *Lysimachia vulgaris* L.
Primrose Family   Primulaceae

Loosestrifes are summer-flowering perennials, mostly 1–3 ft. tall, that grow in moist, shaded or semishaded woods or thickets. Their numerous starlike flowers, each with 5 yellow, pointed petals, may be terminal (clusters at top of the plant), axillary (attached at the leaf nodes), or both.

Garden Loosestrife is a bushy plant with whorled leaves, 3 or 4 per node. Flowers are both axillary and terminal. Native to Eurasia, it escapes from cultivation and grows at lower elevations, mainly in Que. and New Eng.; Jul.–Sep.

Fraser's Loosestrife (*L. fraseri*) is a similar, but rare, native plant of the s. Apps; May, Jun. Swamp Candles (*L. terrestris*) has opposite leaves and numerous flowers in a conspicuous terminal raceme; May–Jul.

Whorled Loosestrife   *Lysimachia quadrifolia* L.
Primrose Family   Primulaceae

Whorled Loosestrife has lanceolate leaves in whorls of 3 to 6. Flowers are both axillary and terminal; note their red centers and long stalks. It occurs in calcareous bogs and moist thickets from ME to GA; May–Aug.

Native Americans made a tea of this plant, which they used to treat kidney, bowel, and other problems.

Whorled Loosestrife should not be confused with Prairie Loosestrife (*L. quadriflora*), which has linear leaves and only axillary flowers; Jul., Aug. Lance-leaved Loosestrife (*L. lanceolata*) has narrow, lanceolate leaves that taper at both ends; Jun.–Aug.

Fringed Loosestrife   *Lysimachia ciliata* L.
Primrose Family   Primulaceae

This loosestrife, similar to Whorled Loosestrife (*above*), can be distinguished from it by its petioles and petals, both of which are fringed. Also, the flowers tend to nod. It is a common plant of sunny swamps and other wet or moist sites throughout the Apps; Jun.–Aug.

Garden Loosestrife
Lysimachia vulgaris

Whorled Loosestrife
Lysimachia quadrifolia

Fringed Loosestrife
Lysimachia ciliata

Yellow Buckeye   *Aesculus flava* Aiton *[A. octandra]*
Buckeye Family   Hippocastanaceae

Buckeyes are deciduous trees bearing opposite, palmately compound leaves with 5–7 pointed leaflets. Their tubular yellow flowers occur in showy panicles that are followed by leathery capsules that contain large, lustrous, brown seeds (buckeyes).

Yellow Buckeye, which may reach 100 ft. or taller, is the largest of our native buckeyes. Its leaves, with leaflets up to 9 in. long, are also the largest. Fruits are smooth, 2 in. across, and contain 1–3 buckeyes. A tree of southern mesophytic forests, its range extends n. to sw. PA.; May, Jun.; fruits, Aug.–Dec.

Two other buckeyes, both native, are smaller trees (40 ft. or less) with yellow flowers: Ohio Buckeye (*A. glabra*), the state tree of that state and seen along the w. edge of the middle Apps, has spiny capsules, May; and Painted Buckeye (*A. sylvatica*) has smooth capsules, Apr., May.

Smooth False Foxglove   *Aureolaria laevigata* (Raf.) Raf.
                                     [Geraradia laevigata]
Figwort Family   Scrophulariaceae

False foxgloves are wiry herbs with bright yellow, bowl-shaped flowers; the 5 equal lobes extend at right angles to the corolla tube. This plant (3 to 5 ft. tall) has smooth, lanceolate leaves with entire margins (the lowest leaves may be lobed or serrate). It occurs in moist to dry deciduous woods from PA and OH southward. Smooth False Foxglove is a hemiparasite on roots of oak trees; Jul., Aug.

Downy False Foxglove (*A. virginica*) has hairy stems and leaves and highly variable leaves. It ranges from s. NH southward; Jul.–Oct.

Big Yellow Wood-sorrel   *Oxalis grandis* Small
Wood-sorrel Family   Oxalidaceae

Wood-sorrels are widespread, low-growing herbs; they have cloverlike leaves, with 3 leaflets that often fold at night and 5-merous flowers.

Big Yellow Wood-sorrel, 1–3 ft. tall, is the largest species of our area; flowers are nearly an inch across. Leaves are often bordered with purple. It is a plant of rich deciduous woods from PA southward; Jun.

Several smaller wood-sorrels with yellow flowers are seen in our mountains. The most common is Yellow Wood-sorrel (*O. stricta*), a cosmopolitan weed; May–Oct.

Yellow Buckeye
Aesculus flava

Smooth False Foxglove
Aureolaria laevigata

Big Yellow Wood-sorrel
Oxalis grandis

Golden Alexanders
Carrot Family

Zizia trifoliata (Michx.) Fern.
Apiaceae

The carrot (or parsley) family is a large temperate one of herbs with flowers in umbels or compound umbels (see fig. 18). Many of the species of our area are Eurasian weeds, but the Appalachian flora includes several native species as well.

In this 1- to 2-ft. plant, each leaf consists of 3 leaflets, each divided into serrate segments as seen here. It occurs in moist woods and along stream banks from Que. southward; Apr.–Jun.

Two other *Zizia* species are also called Golden Alexanders: *Z. aurea*, which is very similar but has leaves with finer teeth, May, Jun.; and *Z. aptera*, with leaves like those of *Z. aureus*, but in addition 1 or more basal, cordate leaves, May, Jun.

Yellow Bedstraw (Gaillet)
Madder Family

Galium verum L.
Rubiaceae

This alien plant, which grows to 2 or 2½ ft., has clusters of fragrant, yellow, 4-petaled flowers. Note the tiny, sharply pointed, linear leaves in whorls. It grows along roadsides and other disturbed places from Que. southward to NC but is much more common in the n. Apps. According to early Christian tradition, this was the bedstraw of Christ's manger in Bethlehem; Jun.–Aug.

White Madder (*G. molluga*) is a similar plant with white flowers; Jun.–Aug. There are several other bedstraws, mostly less showy, weak-stemmed plants with whorled leaves and small, white flowers.

Bush-honeysuckle
Honeysuckle Family

Diervilla sessilifolia Buckley
Caprifoliaceae

Discovered by Samuel B. Buckley in the mid-19th century, Bush-honeysuckle is a shrubby plant 2–6 ft. tall. Note the sessile, lanceolate leaves with serrate margins. The yellowish green flowers are about 2 in. long. Bush-honeysuckle is found in moist soils at elevations from 4,000–6,000 ft. in the s. Apps; Jun., Jul.

Some botanists recognize as variety *rivularis* plants of this species that occur in drier sites and have leaves that are hairy beneath.

Also called Bush-honeysuckle, *D. lonicera* is a more northern species with stalked leaves and flowers that turn, as they age, from yellow to orange to red; Jun., Jul.

Golden Alexanders
Zizia trifoliata

Yellow Bedstraw
Galium verum

Bush-honeysuckle
Diervilla sessilifolia

Bearsfoot   *Polymnia uvedalia* L.
Aster Family   Asteraceae

The aster family (also called the sunflower, daisy, or composite family) is the largest one of the Apps (i.e., has the most species). Members of this family are generally recognized by their characteristic inflorescence: the head, composed of many small flowers (florets) clustered together on a common receptacle. If you examine a single floret under a hand lens, you will see that it has a floral formula of 5-5-5-1, with 5 anthers to form a cylinder around the style; above the style is a 2-lobed stigma. In most species, disc florets comprise the center, with ray florets arranged around the outer edge.

Bearsfoot has hairy, maplelike leaves with winged petioles. The flower heads, 2–3 in. across, include both yellow disc and ray florets. It is a conspicuous plant of mountain meadows and woods from NY southward; Jul., Aug.

Small-flowered Leafcup (*P. canadensis*) has large, toothed, lanceolate leaves and smaller heads with yellow disc florets but white rays (sometimes absent); it is generally a lowland plant of calcareous soils; Jun.–Oct.

Lance-leaved Coreopsis   *Coreopsis lanceolata* L.
Aster Family   Asteraceae

Called tickseeds or tick-seed sunflowers, *Coreopsis* species are herbs with 6–10 (often 8) rays per flower head; rays are yellow and often toothed at their tips. Buds are typically spherical.

This species is a smooth perennial that grows 1 to 2 ft. tall. Leaves are lanceolate, often narrower than those seen here, and usually with 2 basal prongs. It grows naturally in dry soils of open sites from the s. Apps n. to VA and, as a garden escape, to s. New Eng.; May–Jul.

Forest Tick-seed   *Coreopsis major* Walter var. *rigida* (Nutt.) Boynton
Aster Family   Asteraceae

This smooth, 2- to 3-ft.-tall tickseed is recognized by its opposite, paired, sessile leaves, each of which is divided into 3 leaflets, giving the appearance of 6 "leaves" encircling the stem at each node. Heads are 1–2 in. across. It is a common plant in oak forests and other well-drained mountain areas from VA southward. The var. *major* is distinguished from this one by its hairy leaves and stems; Jun.–Aug.

Whorled coreopsis (*C. verticillata*) has leaves divided into filament-like leaflets; Jun., Jul.

Bearsfoot
Polymnia uvedalia

Lance-leaved Coreopsis
Coreopsis lanceolata

Forest Tick-seed
Coreopsis major

Black-eyed-Susan *Rudbeckia hirta* L.
Aster Family Asteraceae

Rudbeckia species, often called coneflowers, have brown, conical discs surrounded by long, yellow rays. Black-eyed-Susan is a bristly, 1- to 3-ft.-tall plant with unlobed leaves and one to several flower heads. Each head, 2 to 3 in. wide, has 10 to 30 rays. The plant brightens roadsides, old fields, and open woods throughout our area; Jun.–Oct.

Thin-leaved Coneflower (*R. triloba*) is a more branched plant with smaller (1 in. across), more numerous heads and 3-lobed lower leaves; Jun.–Oct.

Cut-leaved Coneflower *Rudbeckia laciniata* L.
Aster Family Asteraceae

This tall, highly branched plant has leaves divided into 3–7 sharply pointed lobes (lower leaves are more lobed than the upper ones seen here). Note also the green heads (alternative name, Green-headed Coneflower) and somewhat reflexed rays. It is common from s. Que. southward, typically in moist soils; Jul.–Sep.

Indians made a root tea from several *Rudbeckia* species that was used for worms, indigestion, and snakebite.

Prairie Coneflower (*Ratibida pinnata*) resembles Cut-leaved Coneflower, but it has grayish cones and much more reflexed, lemon yellow rays; May–Aug.

Green-and-gold *Chrysogonum virginianum* L.
Aster Family Asteraceae

The 5 golden, rounded rays and green disc flowers account for the name of this distinctive but infrequently encountered hairy perennial wildflower. Leaves are lanceolate with serrate to dentate margins. Often creeping, and seldom over 1 ft. tall, it is found in moist, shaded woods from s. OH and s. PA s. to GA; Mar.–Jun.

Plants of this species that form mats spread by stolons (horizontal stems) are assigned to the variety *australe,* distinguishing them from those of the more upright variety *virginianum.*

Black-eyed-Susan
Rudbeckia hirta

Cut-leaved Coneflower
Rudbeckia laciniata

Green-and-gold
Chrysogonum virginianum

Grass-leaved Golden-aster
Aster Family

Chrysopsis graminifolia (Michx.) Ell.
Asteraceae

Golden-asters resemble asters but have larger yellow (vs. white or purple) rays. This species, which grows to 3 ft. in height, has silky white stems and soft grasslike leaves that partially curl up. It grows in dry, sandy soils from s. OH and WV southward; Aug.–Oct.

These are also seen infrequently in the middle and s. Apps: Maryland Golden-aster (*C. mariana*), with much broader leaves, Aug.–Oct.; and Prairie Golden-aster (*C. camporum*), which has much smaller leaves, Jul.–Sep.

Giant Sunflower (Soleil)
Aster Family

Helianthus giganteus L.
Asteraceae

Giant Sunflower, which may reach 12 ft. tall, is well-named; its flower heads, though, are only 2 to 3 in. across. Its dentate, lanceolate leaves are, as seen here, alternate, but may be opposite below. It occupies moist sites southward from s. Que.; Aug.–Oct.

Saw-toothed Sunflower (*H. grosseserratus*) has leaves that are stalked and with more prominently toothed margins; Jul.–Oct. These sunflowers have dentate, mostly opposite, lanceolate leaves: Thin-leaved Sunflower (*H. decapetalus*), with thin leaf blades that taper into long petioles and 10 rays per head, Aug.–Oct.; and Woodland Sunflower (*H. divaricatus*), with thick, rough, sessile leaves, Jul.–Sep.

Tick-seed Sunflower
Aster Family

Bidens aristosa (Michx.) Britton
Asteraceae

Bidens aristosa is 2–4 ft. tall and has large, compound leaves, each of which is divided into 5 toothed, sharply pointed leaflets. Flower heads are 1–2 in. wide and have 6–10 rays. This and other *Bidens* species have barbed achenes (small, dry, one-seeded fruits) that attach themselves to fur or clothing. It occurs in sunny, wet sites from New Eng. southward; Aug.–Oct.

Spanish-needles (*B. bipinnata*) is recognized by its wider, incised leaflets and flower heads with shorter, cream-colored rays; Aug.–Oct.

Dwarf Dandelion
Aster Family

Krigia dandelion (L.) Nutt. *[Cynthia dandelion]*
Asteraceae

Krigia species resemble small dandelions. Flower heads are yellow (or golden orange in *K. biflora*) and leaves have jagged margins. Sap is milky. This species, to 1 ft. tall, has solitary flower heads ½ in. across. Note that all leaves are basal. It inhabits open sites from PA and OH s. to VA and TN; Apr.–Jun.

Krigia virginica is very similar, but it typically bears 2 flowers per stalk; Apr.–Aug.

Grass-leaved Golden-aster
Chrysopsis graminifolia

Giant Sunflower
Helianthus giganteus

Tick-seed Sunflower
Bidens aristosa

Dwarf Dandelion
Krigia dandelion

Autumn Sneezeweed   *Helenium autumnale* L.
Aster Family   Asteraceae

This 2- to 5-ft. perennial has winged stems and coarsely toothed, lanceo-late leaves (dried, they cause sneezing). Note the wedge-shaped, slightly reflexed rays that surround the yellow gray globular discs. It grows in swamps and wet meadows throughout the Apps; Aug.–Nov.

Purple-headed Sneezeweed *(H. flexuosum)* has purplish brown discs; Jun.–Oct. Bitterweed *(H. amarum)*, only 1 ft. tall, has narrow, grasslike leaves; Jun.–Oct.

Wingstem   *Verbesina alternifolia* (L.) Britt. *[Actinomeris alternifolia]*
Aster Family   Asteraceae

"Wingstem" refers to the vertical "wings" on stems that are continuous with the leaf petioles. Note the alternate leaves and flower heads with green discs and reflexed rays. The plant, which grows to 7 ft., occurs in moist woods and thickets from the Catskills southward; Aug., Sep.

Yellow Crown-beard *(V. occidentalis)* is very similar but has opposite leaves; Aug.–Oct.

Roundleaf Ragwort   *Senecio obovatus* Muhl.
Aster Family   Asteraceae

"Ragwort" comes from the random arrangement of the yellow rays encir-cling the yellow flower heads, giving them a ragged appearance. Many spe-cies are also called "groundsel." Leaves attached to the flower stalk are typi-cally smaller and more dissected than the basal ones.

The oval basal leaves, broader above the middle, distinguish Roundleaf Ragwort. It grows in woods and on rock outcrops from VT southward; Apr.–Jun.

Golden Ragwort *(S. aureus)* has heart-shaped leaves on long stalks; Apr.–Jul. These ragworts have deeply dissected basal leaves: Squaw-weed *(S. anonymus)*, with solid stems, May, Jun.; and Butterweed *(S. glabellus)*, which has thick, hollow stems, May–Jul.

New England Groundsel (Senecon)   *Senecio schweinitzianus* Nutt.
                                                    [S. robbinsii]
Aster Family   Asteraceae

Also called Robbins'-ragwort, this is a 1- to 3-ft.-tall plant with sharply dentate leaves *(below, center)* that vary from ovate to lanceolate. Primarily a northern species of Que., New Eng., and NY, it also occurs as a disjunct of grassy balds in the high mountains of NC and TN; May–Aug.

Many ragworts, especially *S. aureus,* were used by Indians as diuretics and to treat female ailments.

Autumn Sneezeweed
Helenium autumnale

Wingstem
Verbesina alternifolia

Roundleaf Ragwort
Senecio obovatus

New England Groundsel
Senecio schweinitzianus

Canadian Goldenrod (Verge d'Or)   *Solidago canadensis* L. *[S. altissima]*
Aster Family   Asteraceae

Goldenrods are fall-blooming, perennial herbs with tiny yellow florets in showy clusters. Leaves vary in shape but are always simple and alternate. In general, lowland species have spreading flower clusters like this one; those of higher elevations are usually more wandlike.

Canadian (or Tall) Goldenrod, which may reach 6 ft., has heads clustered on arching side branches that may be 1 ft. long. Its parallel-veined, lanceolate leaves are rough above and hairy beneath; margins are dentate. Stems are grayish, covered with down. Probably our most common goldenrod, this species occurs in both moist and dry open places throughout our area; Aug.–Nov.

Sweet Goldenrod (*S. odora*) is similar but has toothless, anise-scented leaves. Leaves and flowers make a delicious tea; Jul.–Oct.

Contrary to popular belief, goldenrods do not generally cause hay fever. The usual cause, during the time that goldenrods are in flower, are ragweeds (*Ambrosia* species).

Showy Goldenrod   *Solidago speciosa* Nutt.
Aster Family   Asteraceae

Showy Goldenrod is a tall (to 6 ft.) plant with a reddish, thick stem. Leaves toward the base of the stem are large, broad and either entire or weakly toothed. It grows in fields, prairies, and thickets from NH southward; Aug.–Oct.

Stout Goldenrod (*S. squarrosa*) is similar, but the arrangement of its flower heads is more compact and clublike. Basal leaves are also strongly toothed; Aug.–Oct. Large-leaved Goldenrod (*S. macrophylla*) has leaf blades similar to Stout Goldenrod, but bases are rounded, rather than tapered. It is found only in the Catskills and New Eng. northward; Jul.–Sep.

Gray Goldenrod   *Solidago nemoralis* Ait.
Aster Family   Asteraceae

Gray Goldenrod (*S. nemoralis*) is a small (½–2 ft.) plant covered with fine hairs. Note also the tiny stipules (small, paired leaflike structures) at the base of the leaves. It grows in open, dry sites throughout our area; Jul.–Nov.

Bog Goldenrod (*S. uliginosa*) is a larger (2–5 ft. tall) goldenrod with smooth stems and leaves. It prefers moister sites; Sep., Oct.

Canadian Goldenrod
Solidago canadensis

Showy Goldenrod
Solidago speciosa

Gray Goldenrod
Solidago nemoralis

Zigzag Goldenrod *Solidago flexicaulis* L.
Aster Family Asteraceae

Note the zigzag stem of this 1- to 3-ft.-tall goldenrod. The large, broad, toothed leaves account for its also being called Broad-leaved Goldenrod. It thrives in rich forests and thickets from the Gaspé of Que. southward; Jul.–Oct.

Rough-leaved Goldenrod (*S. patula*) also has large, broad, toothed leaves (upper surfaces rough) at the base of the stem, but its flower heads are on long, curving side branches off the main stem; Aug.–Oct.

Erect Goldenrod *Solidago erecta* Pursh
Aster Family Asteraceae

This wandlike goldenrod is a smooth plant that grows to 4 ft. tall. Note the leaves that gradually increase in size downward; those at the base are prominently toothed. Also called Slender Goldenrod, it grows in dry woods and thickets from s. PA southward; Aug.–Oct.

Hairy Goldenrod (*S. hispida*) is similar but has hairy stems and leaves; Sep.–Oct. On alpine summits of New Eng. and the Adirondacks, you may find Alpine Goldenrod (*S. cutleri*); it resembles *S. erecta* but is much shorter (1 ft. or less), has most of its flower heads at the top of the stem, and has a rosette of basal, dentate leaves; Jul.–Sep.

Blue-stemmed Goldenrod *Solidago caesia* L.
Aster Family Asteraceae

The arrangement of the long, lanceolate leaves and spacing of the clusters of flower heads along the bluish stems identify this plant. It occurs in moist woods n. to ME; Aug.–Oct.

Silver-rod *Solidago bicolor* L.
Aster Family Asteraceae

This plant, the only apparently white species of *Solidago* (actually each cluster combines white and yellow florets), is quite unmistakable. The 1- to 3-ft. stem bears ovate, blunt-tipped leaves below the flower heads. White Goldenrod, as it is also called, occurs in dry open woods or rocky places from s. Que southward to NC and TN; Jul.–Oct.

Zigzag Goldenrod
Solidago flexicaulis

Erect Goldenrod
Solidago erecta

Blue-stemmed Goldenrod
Solidago caesia

Silver-rod
Solidago bicolor

Orange

Canada Lily (Lis) *Lilium canadense* L.
Lily Family Liliaceae

True lilies (*Lilium* species) produce several (16–20) large showy flowers in summer; petals are spotted (inside or out). Lanceolate leaves are typically arranged in whorls along the stem.

Canada Lily, which grows to 5 ft. high, is often seen from the Gaspé Peninsula to uplands of VA and KY, invariably in moist woods and meadows; Jun.–Aug.

Plants of this species with brick red flowers (subspecies *editorum*) could be confused with Gray's-lily.

American Indians used a root tea of Canada Lily to treat rheumatism, dysentery, and other ailments. Roots and flower buds can be eaten, and the bulbs can be cooked in soups or stews. One should be cautious, however, because bulbs of some species are poisonous.

Turk's-cap Lily *Lilium superbum* L.
Lily Family Liliaceae

This spectacular plant is distinctive because of its height (6–8 ft.) and the numerous (10–40) flowers on each stem. Each flower has a centrally located "green star." Note the reflexed petals, which form the "Turk's cap." Leaves of the main stem are whorled. It is frequent in bogs, woods, and other moist, partially shaded habitats, especially at high elevations, from ME southward; Jul.–Sep.

Bulbs of Turk's-cap Lily can be eaten as those of Canada Lily (*above*).

Tiger Lily (*L. lancifolium*), which sometimes escapes cultivation, has alternate leaves and dark bulbils (small bulbs) at the nodes; Jul., Aug.

Carolina Lily *Lilium michauxii* Poir. [*L. carolinianum*]
Lily Family Liliaceae

Carolina Lily resembles Turk's-cap Lily (*above*) but is shorter (3 ft. or less), has 3 or fewer flowers per stem, and lacks the "green star." It is also found in drier habitats, particularly oak-pine forests, from VA southward; Aug.

This plant is also called Michaux's-lily in commemoration of the French botanist Andre Michaux (1746–1802).

Canada Lily
Lilium canadense

Turk's-cap Lily
Lilium superbum

Carolina Lily
Lilium michauxii

Day-lily *Hemerocallis fulva* (L.) L.
Lily Family Liliaceae

Unlike true lilies (*Lilium* species, above), day-lilies have long, swordlike, basal leaves and unspotted flowers. This common species often escapes from cultivation and is seen along roadsides and in fields from New Bruns. and New Eng. southward; May–Jul.

Yellow Day-lily (*H. lilioasphodelus*), also from Asia, has fragrant, lemon yellow sepals and petals; it occurs (but less frequently) in much the same habitats; May–Aug.

Flowers of day-lilies can be eaten as fritters or as a seasoning, but some parts of the plant contain toxic compounds. Cultivated hybrid day-lilies come in a wide range of colors and bloom freely from May to September, making them ideal garden perennials.

Blackberry-lily *Belamcanda chinensis* (L.) DC.
Iris Family Iridaceae

Neither a blackberry nor a lily, this tall (2–3 ft.) plant is related to irises, as indicated by its overlapping, swordlike leaves. The flowers suggest those of lilies but have only 3 (vs. 6) stamens. In autumn, the fruits split open, exposing the black seeds, suggesting blackberries. An Asian native, Blackberry-lily has become naturalized throughout much of e. U.S., where it occupies dry, often basic, rocky soils; Jun., Jul.; fruits, Aug.–Oct.

Blackberry-lily is easily established in a wildflower garden by sowing seeds in autumn. Cultivars, with flowers of several colors, are available from nursery and seed companies.

Yellow Fringed Orchid *Habenaria ciliaris* (L.) R. Br. *[Platanthera ciliaris]*
Orchid Family Orchidaceae

The exquisite flowers of this 2-ft.-tall plant vary from yellow to deep orange. Note the long spurs that extend from the lips of each flower. Yellow Fringed Orchid occurs frequently in mountain bogs and other moist habitats from s. New Eng. and NY southward; Jul.–Sep.

The less common Crested Yellow Orchid (*H. cristata*) has smaller flowers with much shorter spurs; Jul., Aug.

These and other members of the orchid family deserve our protection. As they do not transplant well, one should not collect them from the wild.

Day-lily
Hemerocallis fulva

Blackberry-lily
Belamcanda chinensis

Yellow Fringed Orchid
Habenaria ciliaris

Flame Azalea   *Rhododendron calendulaceum* (Michx.) Torr.
Heath Family   Ericaceae

No other Appalachian shrub (or tree) is so unmistakable. William Bartram wrote of Flame Azalea, "in general of the colour of the finest red, orange and bright gold, as well as yellow and cream colour; these various splendid colours are not only in separate plants, but frequently all the varieties and shades are seen in separate branches on the same plant; . . . this is certainly the most gay and brilliant flowering shrub yet known." The erect (to 18 ft.), deciduous shrub produces its 2- to 4-in. lanceolate leaves as the flowers wither. It occurs at low elevations but is most conspicuous in high-elevation balds such as on Roan Mtn., as seen here. Its range extends n. to sw. PA, WV, and se. OH; Apr.–Jun.

The less common Cumberland Azalea (*R. bakeri*) also produces bright yellow/orange/red (usually red) flowers, but they appear after the leaves have expanded; Jun., Jul.

Pale Jewelweed (Impatiente Pale)   *Impatiens pallida* Meerb.
Spotted Jewelweed (Impatiente du Cap)   *I. capensis* Nutt.
Touch-me-not Family   Balsaminaceae

Spurred flowers of jewelweeds hang as if waiting to be pollinated by visiting bees. Pale Jewelweed has yellow flowers (*above*) that contrast with the spotted orange ones of Spotted Jewelweed. "Jewelweed" refers to the colors produced when light strikes drops of water on the dark green leaves. Another name, "Touch-me-not," refers to the ripe fruits, which burst open when touched. Both species are 3- to 5-ft.-tall plants that grow along shaded stream banks throughout our area; Jun.–Sep.

The juice of the leaves of jewelweeds, when applied soon after exposure to Poison Ivy, is helpful in preventing a rash.

Butterfly Weed   *Asclepias tuberosa* L.
Milkweed Family   Asclepiadaceae

This milkweed differs from other *Asclepias* species both by its bright orange (sometimes red or yellow) flowers and the absence of a milky sap. Butterfly Weed grows in dry, sunny sites from s. New Eng. and NY southward; May–Sep.

The name Pleurisy-root indicates its use in the treatment of various respiratory ailments. It can, however, be toxic in improper doses.

Other milkweeds are featured in the pink and the white sections.

Flame Azalea
Rhododendron calendulaceum

Pale Jewelweed   **Spotted Jewelweed**
Impatiens pallida   *Impatiens capensis*

Butterfly Weed
Asclepias tuberosa

Trumpet Creeper   *Campsis radicans* (L.) Seem.
Bignonia Family   Bignoniaceae

This vine may climb 40–50 ft. up a tree and develop a woody stem 2–3 inches in diameter. Note the compound leaves, each with 7–11 sharply lobed, toothed leaflets. Long (3 in.), trumpetlike flowers are adapted for pollination by hummingbirds. The fruit is a seed capsule that reaches 6 in. in length. It occurs in Appalachian woods and thickets n. to MA; Jul.–Sep.

It is sometimes cultivated but can become weedy. "Cow-itch vine" alludes to the skin rash some people get from contact with the plant.

Cross-vine *(Bignonia capreolata)*, of the same family, is distinguished by its simple leaves and coral/cream flowers; Apr.–Jun.

Indian Paint-brush   *Castilleja coccinea* (L.) Spreng.
Figwort Family   Scrophulariaceae

The orange- or scarlet-tipped (rarely yellow or white) 3-lobed bracts are the showy parts of this 1- to 2-ft.-plant. The inconspicuous yellow flowers are all but hidden, except for their protruding pistils. This, the only widespread *Castilleja* species of the Apps, is found in moist, grassy, open places from s. NH southward; Apr.–Aug.

American Indians considered a leaf tea to be useful for rheumatism and other maladies.

North of the range of *C. coccinea* and on high New Eng. mountain peaks, Pale Painted Cup *(C. septrentrionalis)* occurs. It has purple, lanceolate leaves; its inconspicuous flowers are enclosed by cream-colored bracts; Jun.–Aug.

Orange Hawkweed (Eperviere)   *Hieracium aurantiacum* L.
Aster Family   Asteraceae

Hawkweeds are hairy perennials with flower heads, usually yellow, that resemble those of dandelions. The leaves are quite different, however, varying from narrowly lanceolate to ovate, with either dentate or entire margins.

This species, also called King Devil and Devil's-paintbrush, is the only hawkweed of the Apps with orange flowers (others are yellow). The blunt-tipped, ovate leaves, with entire margins, form a basal rosette. It is an alien plant common along roadsides of New Eng. and Canada, and less common s. to NC; Jun.–Sep.

One yellow-flowered species, Shale-barren Hawkweed *(H. traillii)*, is found in that open habitat; Apr.–Jul.

Trumpet Creeper
Campsis radicans

Indian Paint-brush
Castilleja coccinea

Orange Hawkweed
Hieracium aurantiacum

Red

Red Trillium (Trille Dressé)   *Trillium erectum* L.
Lily Family   Liliaceae

Trilliums, noted for their triangular shape, are quite distinctive. The name suggests the 3 broad leaves, as well as the 3 sepals and 3 petals. Flowers are either stalked or sessile (not stalked). Other trilliums are featured in the white and the yellow sections.

Red Trillium is a stalked plant that grows to 1½ ft. tall. The color and width of the petals vary greatly; especially in the s. Apps, the petals may be white (or less often pink or cream) and are often narrower than those seen here. The white variety (*T. erectum* var. *album*) can be distinguished from *T. grandiflorum* by its dark pistils. Red Trillium is seen in moist deciduous and spruce-fir forests from Gaspésia s. to GA; Apr.–Jun.

The names Squaw Root and Beth (birth) Root allude to its use by American Indians to induce childbirth; a root tea was also used at menopause and as an aphrodisiac. It is now known that the roots contain steroids.

Vasey's Trillium   *Trillium vaseyi* Harbison *[T. erectum* var. *vaseyi]*
Lily Family   Liliaceae

Formerly considered a variety of Red Trillium *(above)*, this species is distinguished from it by its nodding flower. Quite limited in its distribution, it is found in moist mountain woods, most abundantly in the region where NC, SC, TN, and GA adjoin; Apr.–Jun.

George Vasey (1822–93) was an American botanist who was born in England.

Sweet Betsy   *Trillium cuneatum* Raf. *[T. hugeri]*
Lily Family   Liliaceae

The largest of our sessile trilliums, this plant has sepals that vary from green to maroon, and petals that are erect. Note also the strongly mottled leaves. It commonly occurs in large colonies, especially on limestone soils. It is more common in middle TN, but its range extends into the mountains of TN and NC southward; Apr., May.

Sessile Trillium (*T. sessile*) is a somewhat smaller (4–12 in. tall) but otherwise very similar plant. It also favors limestone woods but occurs n. of the range of Sweet Betsy; Mar., Apr.

Lance-leaved Trillium (*T. lancifolia*) is also a sessile red-flowered plant. It differs by having much narrower leaves. Common on alluvial soils, it may be seen in ne. AL and nw. GA, especially near Lookout Mtn.; Apr., May.

Red Trillium
Trillium erectum

Vasey's Trillium
Trillium vaseyi

Sweet Betsy
Trillium cuneatum

Painted Trillium   *Trillium undulatum* Willd.
Lily Family   Liliaceae

This is, no doubt, the most distinctive of Appalachian trilliums. The bright red or purplish red marks at the base of the white petals are its hallmark. *Undulatum* refers to the wavy margins of the petals. It is a plant of high-elevation spruce-fir forests in the s. Apps, becoming more common and occurring at somewhat lower elevations northward to the Gaspé Peninsula of Quebec; Apr.–Jun.

Painted Trillium requires a highly acidic, humus-rich soil and cool summer temperatures, and attempts to cultivate it outside its natural range are rarely successful.

Gray's Lily   *Lilium grayi* Watson
Lily Family   Liliaceae

This lily differs from Canada Lily in being smaller (2–3 ft.), with brick red flowers that are more heavily spotted and sepals and petals that are less reflexed. (Note, however, that Canada Lily may also have brick red flowers.) Gray's Lily is confined to high mountain peaks of the s. Apps; Jun., Jul.

This rare lily was named for the famous Harvard University botanist Asa Gray (1810–88).

Wood Lily (*L. philadelphicum*), found from s. Que. to the uplands of KY and TN, has 1 to few orange to red flowers that turn upward; sepals and petals are spotted inside; Jun., Jul.

Red Iris   *Iris fulva* Ker.
Iris Family   Iridaceae

Also called Copper Iris, the flowers of this 1- to 3-ft. plant are orange red to brownish red (occasionally yellow). More common in the swamps of adjacent lowlands, it sometimes occurs in the s. Apps, where it has apparently escaped from cultivation; Apr., May.

Painted Trillium
Trillium undulatum

Gray's Lily
Lilium grayi

Red Iris
Iris fulva

Pawpaw *Asimina triloba* (L.) Dunal
Custard Apple Family Anonaceae

This small deciduous tree produces pawpaws and typically occurs in large numbers in colonies, or "patches." The flowers (1 to 2 in. across), which appear before the foliage, include 3 green sepals that enclose 6 purple petals in two cycles. Leaves are nearly a foot long; note that the broadest part is near the apex. Pawpaws are found in the Apps from w. NY southward, generally at low to midelevations in moist deciduous woods or riverbottoms; Feb.–May; fruits, Jun.–Aug.

The fruits, which resemble small rounded bananas, are edible when ripe (brownish). Both the leaves and seeds, when crushed, have insecticidal properties.

The less common Dwarf Pawpaw (*A. parviflora*) is a smaller tree with smaller leaves, flowers, and fruits; it occurs from NC southward in pine and oak forests; Apr, May; fruits, Jul., Aug.

Sweet Shrub *Calycanthus floridus* L.
Calycanthus Family Calycanthaceae

This, the only N. Amer. genus of the family, includes shrubs with distinctive reddish or brownish flowers 1½ to 2 in. across. Each flower of Sweet Shrub includes several pigmented bracts surrounding the sepals and petals of the same color. When crushed, the leaves and flowers release a fragrance suggestive of that of strawberries. It occurs infrequently at low elevations of the s. Apps, especially in cove hardwood forests; Apr.–Jun.

The fruits are brown pods that contain seeds often eaten by white-footed mice. Also called Carolina Allspice, the shrub is cultivated, especially in the Appalachians, as a garden ornamental.

Wild Ginger (Asaret du Canada) *Asarum canadense* L.
Birthwort Family Aristolochiaceae

In the axil of each pair of cordate leaves is a solitary flower at ground level. Lacking petals, the 3 sepals, each with a long tip, form the conspicuous maroon calyx tube. (The position of the flowers allows the seeds to be easily reached by ants, which disperse them.) Wild Ginger is scattered in rich woods throughout the Apps; Apr., May.

Native Americans used its aromatic roots to make a tea to treat a wide variety of ailments. Modern research has revealed the presence of aristolochic acid, an antitumor compound. The roots have also been used as a substitute for true ginger, which comes from an unrelated tropical plant.

Pawpaw
Asimina triloba

Sweet Shrub
Calycanthus floridus

Wild Ginger
Asarum canadense

Fire Pink Silene virginica L.
Pink Family Caryophyllaceae

Fire Pink is a 1- to 2-ft.-tall plant with opposite, lanceolate leaves. Note the 5 notched petals at right angles to the corolla tube. It is common in open woodlands, especially semishaded banks, from PA southward; Apr.–Jun.

There is some evidence that American Indians used roots of Fire Pink to get rid of intestinal worms.

The aptly named Round-leaved Catchfly (*S. rotundifolia*), with similar flowers but roundish leaves, is occasionally found on sandstone ledges of the Cumberland Plateau from KY to GA; Jun.–Sep.

Other *Silene* species have white flowers: Starry Campion (*S. stellata*), with fringed petals and whorled leaves, Jun.–Sep.; Bladder Campion (*S. cucubalus*), with opposite leaves and greatly inflated, veined bladders (calyx tubes), May–Aug.

Red Buckeye Aesculus pavia L.
Buckeye Family Hippocastanaceae

Compared to Yellow Buckeye, this is a much smaller (to 25 ft.) tree with smaller and narrower leaflets. Its red flowers produce smaller, but also smooth, capsules with 1–2 buckeyes. Although primarily a Coastal Plain species, it occurs sporadically in the s. Apps, where it may be locally abundant on moist sites as far n. as s. WV; Apr., May; fruits, Aug.–Dec.

It is sometimes cultivated as an ornamental.

Flowering Raspberry (Chapeaux Rouges) Rubus odoratus L.
Rose Family Rosaceae

Flowering Raspberry is unique among *Rubus* species because of its spineless branches, maplelike leaves, and rose purplish flowers. The purple fruits are edible, but they taste dry and insipid. It occupies the edges of woods throughout the mountains; May–Sep.; fruits, Jul.–Sep.

Other *Rubus* species—raspberries, dewberries, and blackberries, or collectively, "brambles"—have prickly canes, compound leaves, and white flowers.

A tea made from the leaves and roots of various brambles was widely used by American Indians for "female problems" and other ailments.

Fire Pink
Silene virginica

Red Buckeye
Aesculus pavia

Flowering Raspberry
Rubus odoratus

Cardinal Flower   *Lobelia cardinalis* L.
Bluebell Family   Campanulaceae

Leaves of Cardinal Flower are alternate, lanceolate, and dentate. This, our only red lobelia, grows 2 to 3 ft. tall along the shores of streams and lakes from s. Que. and New Bruns. southward; Jul.–Sep.

Cardinal Flower was used medicinally by Indians, much as were the blue lobelias (blue/purple section) but less often. A love potion was made from its roots.

Common Lousewort   *Pedicularis canadensis* L.
Figwort Family   Scrophulariaceae

This low-growing (5–10 in.) perennial has deeply incised leaves, most of which are basal, but a few occur on the flower stalk. Hooded flowers may be yellow or red or bicolored red/yellow, as seen here. Also called Wood-betony, it is found in open woods from s. Que. southward; Apr.–Jun.

Both "Lousewort" and *Pedicularis* indicate its former use as an insecticide, especially for lice. American Indians, who considered the roots to be an aphrodisiac, also used a root tea for digestive and heart ailments.

The less common Swamp Lousewort (*P. lanceolata*) is a taller, fall-blooming plant of wet sites; Aug., Sep.

Crimson Bee Balm   *Monarda didyma* L.
Mint Family   Lamiaceae

This, the showiest of the bee balms, has crimson tubular flowers crowded on a head above dark red bracts. Other features of this 2- to 3-ft.-tall plant are typical of the family: square stems, opposite leaves, and a minty odor when crushed. It is a plant of moist woods from the s. Apps n. to NY and, escaping from cultivation, to Que.; Jul.–Sep.

Indians used a leaf tea to treat a long list of ailments. A refreshing hot tea is also made from the flower heads. Such uses account for its also being called Oswego tea.

Purple Bergamot (*M.* × *media*), believed to be a hybrid between *M. didyma* and another *Monarda* species, is very similar but has reddish purple flowers, Jul.–Sep.; Horsemint (*M. punctata*) has tiered whorls of yellow flowers punctuated by purple spots, Jul.–Sep.

Cardinal Flower
Lobelia cardinalis

Common Lousewort
Pedicularis canadensis

Crimson Bee Balm
Monarda didyma

Spice-bush
Laurel Family

Lindera benzoin (L.) Blume
Lauraceae

This highly branched shrub grows 6 to 12 ft. tall; leaves are 2 to 5 in. long. Its tiny, yellowish flowers are arranged in clusters less than ½ in. across. Staminate and pistillate flowers are borne on separate plants. One of the earliest shrubs to flower, it is commonly seen along stream banks and other moist to wet sites at low elevations from GA to s. ME; Mar.–May; fruits, Aug.–winter.

The name Wild Allspice also indicates the aromatic properties of its stems and leaves.

Sassafras (*Sassafras albidum*) of the same family is also an aromatic plant with small, white flowers that appear before the leaves. A tree that may become more than 100 ft. tall, its leaves are of several shapes: unlobed, "left-handed mittens," "right-handed mittens," and 3-lobed. The bark is used to brew sassafras tea, long used in the Appalachians as a spring tonic; Mar.–Jun.

Pokeweed
Pokeweed Family

Phytolacca americana L.
Phytolaccaceae

This distinctive, coarse perennial may grow to 10 ft. Both the stems and veins of the lanceolate leaves are noticeably dark red. Seen here are green berries and dark purple ripe ones. Earlier, the small, numerous, white flowers were in racemes. Poke is common in waste places and fencerows from New Eng. southward; May–Oct.; fruits, Aug.–Nov.

Eating "poke salat" as a green or potherb is an old Appalachian tradition. However, careful preparation of new spring leaves, including parboiling, is necessary to prevent poisoning; the older stems, roots, and berries have the highest concentration of toxins. Juice from the berries was used in colonial days as an ink.

Strawberry Bush
Stafftree Family

Euonymus americana L.
Celastraceae

Fruits of this erect or arching shrub are more conspicuous than its small, greenish yellow, 5-petaled flowers. The pink fruits break open to expose its orange red seeds. The opposite leaves are widest near the middle. Often called Hearts-a-busting-with-love by mountain folk, it is common in open woods and ravines north to PA; May; fruits, Sep., Oct.

Two other shrubby *Euonymus* species with showy fruits are Wahoo (*E. atropurpureus*), which has clusters of purplish flowers followed by smooth, rose-colored, 4-lobed fruits, May, Jun.; and Trailing Strawberry-bush (*E. obovatus*), a prostrate shrub or trailing vine that forms knobby 3-lobed fruits, May, Jun.

Spice-bush
Lindera benzoin

Pokeweed
Phytolacca americana

Strawberry Bush
Euonymus americana

Fire Cherry (Cerises d'été)   *Prunus pensylvanica* L.f.
Rose Family   Rosaceae

Prunus includes such familiar trees as peach, plum, and cherry. This small tree has long (3–4 in.), serrate, lanceolate leaves attached to reddish twigs. Flowers are ½ in. wide, have white rounded petals, and are borne on stalks of unequal lengths. Fruits, ¼ in. in diameter, are eaten by wildlife and can also be used to make a tasty jelly. Seedlings of Fire Cherry are rapidly established after a forest fire, explaining their common occurrence in nearly pure stands. It is common across s. Canada, with a range extension down the Apps to NC and TN; Apr.–Jul.; fruits, Jul.–Sep.

Chickasaw Plum (*P. angustifolia*) is a similar small tree, but it has smaller leaves (1½–2 in. long) and sepals that are hairy underneath; Apr. Black Cherry (*P. serotina*), is a much larger tree (to 80 ft.); its flowers are similar but are arranged in elongated, drooping racemes. Fruits are dark red to almost black and are inedible; May.

Blackberry   *Rubus* species
Rose Family   Rosaceae

Blackberries have erect or arching stems. Their compound leaves generally have 3 or 5 leaflets. Flowers are about ½ in. across and produce red fruits that turn black as they ripen. Identification of blackberries to species requires a specialist. Blackberries and other *Rubus* species—collectively "brambles"—are found in open disturbed sites throughout the Apps; May–Jul; fruits, Jul.–Sep.

Thorns, seen here on a low-elevation specimen, are absent on those at higher elevations.

Wild Strawberry   *Fragaria virginiana* Duchesne
Rose Family   Rosaceae

This low-growing native plant (less than 1 ft. tall) has typically rosaceous flowers with 5 white petals and numerous stamens. Note the 3 toothed leaflets. Fruits, ½ in. in diameter, are smaller than cultivated strawberries but are considered tastier; Apr.–Jun.; fruits, Jun.–Sep.

The cultivated strawberry (*F. ananassa*) is derived from a cross of *F. virginiana* with *F. chiloensis*, a species native to Chile.

Indian Strawberry (*Duchesnea indica*) has similar, but yellow-petaled, flowers; its red fruits are tasteless; Apr.–Jul.; fruits, Jun.–Sep.

Fire Cherry
Prunus pensylvanica

Blackberry
Rubus

Wild Strawberry
Fragaria virginiana

Smooth Sumac (Vinaigrier)   *Rhus glabra* L.
Cashew Family   Anacardiaceae

Sumacs are deciduous shrubs or small trees of roadsides and open fields. Their compound leaves have historically served as a source of tannins for leather, apparently explaining the name "Shoemake" that is sometimes applied. They have clusters of tiny, greenish yellow flowers that become clusters of hairy, red fruits; the fruits can be used to make a refreshing cold drink that tastes much like lemonade. All four sumac species described here are found throughout most of our area.

Smooth Sumac is noted for its smooth twigs and petioles. The large leaves are divided into 11–31 dentate, lanceolate leaflets. Small white or green flowers in a panicle up to 8 in. long are followed by the sharply pointed cluster of hairy fruits seen here; Jun.–Aug.; fruits, Sep.–winter.

Staghorn Sumac (*R. typhina*) differs primarily by having hairy branches and petioles; Jun.–Aug.; fruits, Aug.–winter.

Winged Sumac   *Rhus copallinum* L.
Cashew Family   Anacardiaceae

Winged Sumac is similar to Smooth Sumac (*above*), but can be easily distinguished by its prominent wings that extend along each side of the rachis, the stalk between the leaflets. Its leaflets are also glossier, accounting for its other name, Shining Sumac. Note that its fruits are not as brightly red as those of Smooth Sumac; Jul.–Sep.; fruits, Sep.–Feb.

Fragrant Sumac   *Rhus aromatica* Ait.
Cashew Family   Anacardiaceae

Several features distinguish this aromatic sumac from the ones above. Its leaves are, as seen, trifoliate. The small yellow flowers, which appear before the leaves, are in much smaller clusters and on side branches. Fruits (berries) are larger, mature earlier, and are in clusters of a dozen or fewer. It invariably inhabits dry, calcareous soils; Apr.; fruits, Jun., Jul.

Poison Ivy (*Toxicodendron radicans*) is a shrubby vine with variable leaves that may resemble those of Fragrant Sumac. A distinguishing feature is the prominent petiole of each terminal leaflet of Poison Ivy. Also its fruits (berries) are white; May, Jun.; fruits, Jul.–Nov. Unlike Poison Ivy, Fragrant Sumac does not cause a skin rash.

Smooth Sumac
Rhus glabra

Winged Sumac
Rhus copallinum

Fragrant Sumac
Rhus aromatica

American Holly   *Ilex opaca* Ait.
Holly Family   Aquifoliaceae

Hollies are evergreen or deciduous shrubs or small trees. Small staminate (male) or pistillate (female) flowers are typically borne on separate plants (i.e., hollies are dioecious). Fruits are red to dark red berries.

Ilex opaca is the only evergreen Appalachian holly. Leaves of the medium-size tree (to 40 ft.) are leathery and prickly; bark is smooth and dark gray. Small flowers are followed by bright red berries, making holly branches a favorite Christmas decoration. Primarily a tree of southeastern lowland forests, it occurs at low to middle elevations of the Apps only as far north as WV; Apr.–Jun.; fruits, Jun.–winter.

Besides its decorative uses, the close-grained, white wood of American Holly is in demand for such items as ship models and furniture inlays.

Winterberry (Houx Verticillé)   *Ilex verticillata* (L.) A. Gray
Holly Family   Aquifoliaceae

This deciduous shrub or small tree (to 25 ft.) has leaves 2–4 in. long, often as seen here but sometimes with toothed margins. Small white flowers and, later, red berries are on short (less than ½ in.) stalks. Except for Gaspésia, it is found throughout Appalachia on moist to wet sites; Jun.; fruits, Oct.–Jan.

The principal other deciduous Appalachian holly is Mountain Holly (*I. montana*). It has larger (4–6 in. long) leaves with serrate margins and flowers/berries on longer (more than ½ in.) stalks; May, Jun.; fruits, Oct.–Jan.

Carolina Buckthorn   *Rhamnus caroliniana* Walt.
Buckthorn Family   Rhamnaceae

Buckthorns are shrubs or small trees with pinnately veined simple leaves; small clusters of axillary greenish white flowers are followed by "berries" (actually drupes) that are black when mature. Many buckthorns have large paired spines on their branches.

Carolina Buckthorn, a thornless species, grows to 25 ft. Its glossy leaves, 2–5 in. long, have finely serrate margins. The berries, red in summer, turn black in fall. Also called Indian Cherry, it is primarily a southern plant of rich woods and roadsides that occurs from VA, WV, and OH southward; May, Jun.; fruits, Sep.–Nov.

Several other *Rhamnus* species are more widely distributed in the Apps: Common Buckthorn (*R. cathartica*), a thorned European naturalized shrub, Apr., May; and Lance-leaved Buckthorn (*R. lanceolata*), a tall unarmed (thornless), native shrub, Apr., May.

American Holly
Ilex opaca

Winterberry
Ilex verticillata

Carolina Buckthorn
Rhamnus caroliniana

American Ginseng (Ginseng à Cinq Folioles)   *Panax quinquefolius* L.
Ginseng Family   Araliaceae

This 1- to 2-ft. perennial has compound leaves with 3–5 leaflets and small, white flowers in showy, spherical umbels. Originally, it was common and widespread throughout the Apps from Que. to GA. Due to extensive collection, however, "sang" is now in need of protection; Jun., Jul.; fruits, Aug.–Oct.

Due to its irregularly shaped roots, roughly suggestive of a man's body, ginseng has long been highly regarded by Asians (and Indians to a lesser extent) as a panacea and aphrodisiac. Modern tests by Western scientists reveal the presence of several physiologically active compounds that are consistent with these uses.

Dwarf Ginseng (*P. trifolius*) is a smaller plant with an onion-shaped root and leaves with 3 narrow leaflets. It has had limited medicinal use by Amer. Indians, but it is not of commercial importance; Apr., May.

Partridge-berry (Pain de Perdrix)   *Mitchella repens* L.
Madder Family   Rubiaceae

Unmistakable when in either flower or fruit, this creeping vine has small, ovate, evergreen leaves. At the end of the stem is a pair of white (or pinkish) tubular flowers, each with 4 lobes. From each pair of flowers is produced a single red fruit (edible, but dry and tasteless) that persists through the winter. Partridge-berry is found in shaded slopes in woods from s. Que. southward. The only other species of this genus is native to Japan; May–Jul.; fruits, Aug.–Mar.

An alternative name, Squaw-vine, recognizes the use of a leaf or berry tea made by Indian women for such problems as childbirth pain, painful menses, and sore nipples.

Red Elderberry (Sureau Rouge)   *Sambucus racemosa* L.
Honeysuckle Family   Caprifoliaceae

Elders are shrubby plants with opposite, pinnately compound, deciduous leaves. Small flowers are clustered much like those of viburnums. Fruits are juicy berries with 3 small seeds.

Red Elderberry, 25 ft. tall, has 5–7 leaflets per leaf, and white or cream flowers in rounded clusters. Fruits (berries) are occasionally white or yellow. It is found sporadically throughout the Apps, at higher elevations in the south, at lower ones in the north; Apr.–Jul.; fruits, Jun.–Aug.

Common Elderberry (*S. canadensis*), a lowland species, has white flowers in flat-topped cymes, followed by purplish black berries.

American Ginseng
Panax quinquefolius

Partridge-berry
Mitchella repens

Red Elderberry
Sambucus racemosa

Pink

Rosy Twisted Stalk (Rognons de Coq) *Streptopus roseus* Michx.
Lily Family Liliaceae

Small, pink- or rose-colored, bell-like flowers are arranged along the 1- to 2-ft. zigzag stems of this branching plant. Its twisted flower stalks reflect both the common and genus (*Streptopus*) name. Shade-tolerant, it is found in rich woods at high elevations throughout the Apps; May–Jul.

White Mandarin (*S. amplexifolius*) has flowers that vary from green to purple but that are typically white. It is infrequent in the s. Apps, becoming more common northward in the spruce-fir forests of Canada; Apr.–Aug.

Nodding Onion *Allium cernuum* Roth.
Lily Family Liliaceae

The crook at the top of the flower stalk (1 ft.) makes this plant distinct from other *Allium* species. The grasslike leaves are shorter than the flower stalk. Nodding Onion is found on ledges, rocky soils, and open woods, often at high elevations, from the Catskills southward; Jul., Aug.

Some botanists assign the similar onion with fewer, nearly white flowers that is found in shale barrens to the species *A. oxyphilum*; Jul., Aug. The Ramp (*A. tricoccum*), or Ail des Bois, produces broad, lanceolate leaves in spring, followed in summer by a cluster of white flowers. Both the leaves and bulbs are edible but have a very strong garliclike odor; Jun., Jul.

These and other *Allium* species have long been used medicinally for a variety of ailments. Modern research has confirmed in many cases their efficacy.

Swamp Pink *Helonias bullata* L.
Lily Family Liliaceae

Perhaps no other Appalachian plant would be mistaken for this one. Above a rosette of 8- to 12-in.-long lanceolate leaves extends a hollow, 1–3 ft., leafless stem that bears an oval cluster of bright pink flowers. As indicated by its common name, it is found in swamps and bogs. It is primarily a Coastal Plain species, but it occurs in widely separated populations in the mountains from PA southward; Apr., May.

Swamp Pink is threatened due to the destruction of bogs and its own attractiveness, which encourages its being casually picked. Its rarity and uniqueness make it an ideal candidate for special conservation measures.

Rosy Twisted Stalk
Streptopus roseus

Nodding Onion
Allium cernuum

Swamp Pink
Helonias bullata

Pink Moccasin Flower (Cypripede Acaule) *Cypripedium acaule* Ait.
Orchid Family Orchidaceae

This handsome plant bears pink flowers on stalks separate from the leaves. It occupies a variety of habitats, ranging from relatively dry soils of oak-pine forests to bogs. It is found throughout the mountains, where it may be locally abundant; Apr.–Jul.

Both this plant and Yellow Lady's-slipper have been called American Valerian because of their use, like the unrelated European Valerian (*Valeriana officinalis*), as a sedative.

Many of our orchids are rare and endangered, and thus afforded government protection. The natural urge to transplant them should be resisted as such efforts are rarely successful.

Showy Lady's-slipper (Cypripède Royal) *Cypripedium reginae* (Walt.)
Orchid Family Orchidaceae

Also called Queen Lady's-slipper, this 1- to 3-ft.-tall orchid typically bears two rose-and-white flowers per flower stalk. It is found sporadically in moist to wet sites throughout the Appalachians but is never abundant, partly due to digging by collectors. Contact with the hairy leaves can cause an eczema; May, Jun.

The Small White Lady's-slipper (*C. candidum*) is a smaller plant with an all-white slipper. Primarily an upper midwestern species, it is found along the western edge of the Appalachians from NY to KY; May, Jun.

Showy Orchis *Orchis spectabilis* L. *[Galearis spectabilis]*
Orchid Family Orchidaceae

This orchid has small (½ in. wide) purple-and-white flowers. Note the pair of leaves at the base of the 6–12 in. stalk. It is found in rich woods from Que. to GA; Apr.–Jun.

One-leaved Orchis (*O. rotundifolia*) is a much less common but slightly larger plant that is restricted to the n. Apps and adjacent lowlands. It has a solitary, round leaf attached to the flower stalk; Jun., Jul.

Pink Moccasin Flower
Cypripedium acaule

Showy Lady's-slipper
Cypripedium reginae

Showy Orchis
Orchis spectabilis

Rose Pogonia *Pogonia ophioglossoides* (L.) Ker Gawler
Orchid Family Orchidaceae

Each flower (1 per stem) has 3 sepals and 2 petals, all of which are similar. The third, crested, highly fringed petal is distinctive. In addition to the upper leaf seen here, another is attached at midstem. This 4- to 18-in. plant is found in mountain bogs and wet, acidic meadows from Que. s. to TN; Jun.–Aug.

Found in much the same habitats and range is Arethusa, or Dragon's-mouth, *(Arethusa bulbosa)*; it's a similar but smaller plant. It has 3 pink, erect sepals; its conspicuous petal is yellow-crested and not so highly fringed; May, Jun.

Lily-leaved Twayblade *Liparis lilifolia* (L.) Rich
Orchid Family Orchidaceae

The 4- to 10-in.-long raceme of this plant includes 12–24 delicate, semi-transparent mauve flowers. Note the pair of glossy basal leaves (2–5 in. long) that account for its common name. It occurs infrequently along stream and pond banks from s. NH southward; May–Jul.

The less common Yellow Twayblade, or Fen-orchid, *(L. loeselii)* is a smaller plant with smaller, yellow green flowers. It grows in bogs and seeps from the Gaspé southward; Jun., Jul.

Spotted Coral-root *Corallorhiza maculata* (Raf.) Raf.
Orchid Family Orchidaceae

This is one of a dozen or so saprophytic orchids of N. Amer; their branching rhizomes are connected to filaments of soil fungi, from which they derive nourishment. This large (1 to 2 ft. tall) and rather common species has purple flowers with purple spots on white ruffled lips. Note the reddish stems. It is found widely throughout the Apps in both coniferous and deciduous forests; Apr.–Sep.

Spring Coral-root *(C. wisteriana)* is a similar but smaller and less common orchid that blooms earlier; Apr., May. Autumn Coral-root *(C. odontorhiza)* has more green in its stem and flowers, and flowers later; Aug.–Oct.

The roots of various *Corallorhiza* species have long been used as a folk remedy for fevers.

Rose Pogonia
Pogonia ophioglossoides

Lily-leaved Twayblade
Liparis lilifolia

Spotted Coral-root
Corallorhiza maculata

Leatherflower   *Clematis viorna* L.
Buttercup Family   Ranunculaceae

Each of the urn-shaped flowers (1 in. long) of this vine is enclosed by 5 thick, pink, petal-like sepals (petals are absent). The leaves are divided into 3–7 leaflets. It is found in the mountains from PA and OH southward; May–Aug.

Other *Clematis* species with solitary flowers include Purple Clematis (*C. occidentalis*) with mauve sepals and leaves with 3 leaflets; Apr.–Jun. At least two species are limited to shale barrens of WV or VA: *C. viticaulis*, with green sepals, Apr.–Jun; and *C. albicoma*, with purple ones, Apr.–Jun.

Virgin's-bower (*C. virginiana*), or Herbe aux Geux, a common species throughout e. N. Amer., is a vine with creamy white flowers in panicles followed by fruits with gray, silky plumes; Jul.–Sep.; fruits, Sep., Oct.

Wild Columbine (Gants de Notre-Dame)   *Aquilegia canadensis* L.
Buttercup Family   Ranunculaceae

This is the only native columbine of e. N. Amer. Each of the graceful inverted flowers has 5 spurs with nectar inside, which attracts hummingbirds. Inside are sepals, usually cream. Leaves are compound with 3 leaflets. It is common throughout our area, especially on dry, partially shaded banks; Apr.–Jul.

Native Americans used crushed seeds to treat headaches and to control lice; roots were chewed to correct various digestive problems.

Wild Bleeding-heart   *Dicentra eximia* (Ker Gawler) Torr.
Fumitory Family   Fumariaceae

As compared to the more common Dutchman's-breeches, flowers of *D. eximia* are somewhat longer and a deep pink (or sometimes purple). Also, they are in panicles, rather than racemes. Note the deeply dissected leaves. Wild Bleeding-heart is seen on rocky ledges or woods, sometimes along streams and often at high elevations. Its range extends from w. NY to TN and GA; May–Aug.

The garden variety of this species, which bears larger and showier pink-and-white flowers on arching stems, sometimes escapes from cultivation.

Leatherflower
Clematis viorna

Wild Columbine
Aquilegia canadensis

Wild Bleeding-heart
Dicentra eximia

Bouncing-bet *Saponaria officinalis* L.
Pink Family Caryophyllaceae

This stout perennial has numerous white to pink flowers, each of which consists of a corolla tube and 5 slightly notched, reflexed petals. The lanceolate leaves are wider toward their bases. It grows along roadsides, railroad tracks, and other waste places throughout our area; May–Oct.

Rub moistened leaves between your hands and the lather formed explains an alternative name, Soapwort. Bouncing-bet is an old term for a woman who washes clothes; thus both common names refer to the use of the plant as a soap substitute. The same saponins responsible for lathering also can cause severe digestive problems if the leaves are ingested.

Wild Pink *Silene caroliniana* Walt.
Pink Family Caryophyllaceae

This 6- to 10-in.-tall mountain wildflower is recognized by its unnotched or barely notched, wedge-shaped petals and sticky stems and sepals. Petals vary in color from white to deep pink. Wild Pink occurs in dry, calcareous woods and openings from s. NH southward to n. AL; Apr.–Jun.

Northern Pitcher Plant (Sarracénie Pourpre) *Sarracenia purpurea* L.
Pitcher Plant Family Sarraceniaceae

Plants with leaves modified into insect-trapping pitchers that fill with water characterize the genus *Sarracenia*. Like other carnivorous plants, they grow in open, sunlit, boggy sites. Protein absorbed from digested insects compensates for the low nitrogen content of the soil.

In this species, several heavily veined pitchers, leaning outward, surround the stalk that bears the round, pink to reddish purple nodding flower. The flower is held a foot or so above the pitchers, an arrangement that prevents potential insect pollinators from also being trapped. It is an infrequent plant of mountain bogs from s. Canada southward; Apr.–Jun.

Indians of Canada used this plant as a preventative for smallpox, but its efficacy has not been tested.

Sweet Pitcher Plant (*S. jonesii*) is similar but has erect pitchers; May, Jun.

Bouncing-bet
Saponaria officinalis

Wild Pink
Silene caroliniana

Northern Pitcher Plant
Sarracenia purpurea

Pink Knotweed *Polygonum pensylvanicum* L.

Buckwheat Family Polygonaceae

Many *Polygonum* species are called "knotweed" because of a sheath surrounding the nodes. The name "smartweed" is also applied to many of these same plants because of the sharp taste of their (simple) leaves. They have very small, usually pink, petal-like sepals in spikes or spikelike racemes.

Also called Pennsylvania Smartweed, this erect, branching annual grows to 5 ft. in height. It is found in open places, especially in rich, moist soil, throughout e. N. Amer.; May–Oct.

Water Smartweed (*P. amphibium*) is a highly variable plant, also with pink flowers but with larger leaves; it is found in and along the margins of ponds and swamps; Jul.–Sep.

Alpine Bistort (*P. viviparum*) is a circumboreal plant common in the alpine areas of Mt. Washington and Mt. Katahdin. Only 4 to 12 in. tall, the plant has pink flowers; stem leaves are narrowly lanceolate, but basal leaves are ovate and long-petioled; Jun.–Aug.

Swamp Rose-mallow *Hibiscus moscheutos* L.

Mallow Family Malvaceae

Members of this family, which include cotton, okra, and hollyhock, are recognized by numerous stamens that surround and are attached to the style.

This tall (5–7 ft.) perennial, with its hollyhock-like flowers, is quite spectacular. If pink, the petals may or may not have a dark center; if white, the petals always have a dark center. Swamp Rose-mallow is primarily a plant of coastal swamps, but it occurs infrequently in mountain marshes n. to WV and VA; Jul.–Sep.

Halberd-leaved Rose-mallow (*H. laevis*) has large pink flowers, but its leaves have 3 sharply pointed lobes; Aug., Sep.

Musk Mallow *Malva moschata* L.

Mallow Family Malvaceae

This alien, 1- to 2-ft.-tall plant has intricately cut musk-scented leaves; the 5 pink (or white) petals are notched. It is a weedy wildflower that occurs throughout Appalachian roadsides, becoming less common s. to VA; Jun., Jul.

Two other mallows are also roadside weeds with 5-petaled pink flowers: Common Mallow (*M. neglecta*), a small creeping plant with roundish leaves and slightly notched petals, May–Oct.; and High Mallow (*M. sylvestris*), which more closely resembles Musk Mallow but has 5-lobed leaves and petals with prominent, dark red veins, Jun.–Aug. Both occur throughout the Apps.

Pink Knotweed
Polygonum pensylvanicum

Swamp Rose-mallow
Hibiscus moscheutos

Musk Mallow
Malva moschata

Spring-beauty (Claytonia de Caroline)   *Claytonia virginica* L.
Purslane Family   Portulacaceae

Like other members of the purslane family, Spring-beauty has flowers with 2 sepals, 5 petals, and 5 stamens. Note the pink-veined white petals and linear leaves. As it propagates itself by corms (root swellings), it often occurs in large colonies. It is commonly found in rich woods throughout the Appalachians; Mar.–May.

Carolina Spring-beauty *(C. caroliniana)*, recognized by its broad leaves and distinct petioles, occupies very similar sites but commonly occurs at higher elevations; Mar.–May.

Corms of both species are sometimes boiled and eaten like potatoes; for conservation reasons, this is not recommended except in an emergency.

Allegheny Stonecrop   *Sedum telephoides* Michx.
Orpine Family   Crassulaceae

This stonecrop is easily recognized by its large size (1 ft. tall) and wide, fleshy leaves attached to a thick stem; the dense flower clusters are 2–4 in. across. The entire plant is often tinged with pink or purple. It is a common Appalachian plant of middle and higher elevations, growing on cliffsides from w. NY to GA. Other names are Wild Live-forever and American Orpine; Aug., Sep.

The similar Live-forever, or Orpine *(S. purpureum)*, is a garden escape that occurs in the n. Apps. It has pink or purple flowers and leaves with coarsely dentate teeth. Several parts of the plant are edible. The young leaves may be eaten in a salad or boiled as a vegetable. The large tubers can be boiled like potatoes or pickled in vinegar; Jul.–Sep.

Lime Stonecrop   *Sedum pulchellum* Michx.
Orpine Family   Crassulaceae

The tiny, white to pink flowers of this plant are arranged in inflorescences with 3–7 branches. The leaves ($\frac{1}{2}$ in. across) are narrow and cylindrical. More common in rocky soil west and south of the Appalachians, Lime Stonecrop occurs, though infrequently, along gravely mountain roadsides throughout most of the Apps; May, Jun.

These garden escapes have yellow flowers: Mossy Stonecrop *(S. acre),* with pointed, overlapping leaves, Jun., Jul.; and Stonecrop *(S. sarmentosum)* with its longer, whorled leaves in 3s, Jun.

Spring-beauty
Claytonia virginica

Allegheny Stonecrop
Sedum telephoides

Lime Stonecrop
Sedum pulchellum

Goat's-rue   *Tephrosia virginiana* (L.) Pers.
Pea Family   Fabaceae

Also called Hoary Pea because of its silky hairy covering, this unbranched, 1- to 2-ft.-tall plant has distinctive bicolored (yellow/pink flowers). It is common in sunny places with sandy soils, such as openings in oak-pine forests. Its range in the mountains extends from New Eng. southward, but it is more common to the west; May–Jul.

Southern Goat's-rue *(T. spicata)* is a less common, branched plant. The standard (uppermost petal) of each flower is purple (vs. yellow in *T. virginiana)*; Jun.–Aug.

Tephrosia species are known for their usefulness as insecticides and fish poisons (after being stunned in the water, fish can be caught by hand). Goat's-rue has also been used for the treatment of tuberculosis and coughs. Modern research reveals that it has potential in cancer treatment.

Spurred Butterfly Pea   *Centrosema virginianum* (L.) Benth.
Pea Family   Fabaceae

This twining vine with trifoliate compound leaves can climb to 6 ft. Each narrow leaflet is 1–4 in. long. Note the conspicuous (1–1½ in. long) standard (large spreading upper petal). It occupies sandy soils of open woodlands from VA and KY southward; Jun.–Aug.

Butterfly Pea *(Clitoria mariana)* is also a twining vine and has very similar flowers and compound leaves, but its leaflets are much wider; Jun.–Aug.

Virginia Meadow-beauty   *Rhexia virginiana* L.
Meadow-sweet Family   Melastomataceae

This ft.-tall herb has wide, lanceolate, opposite leaves arranged below its distinctive flowers. Note the 4 petals that surround the 8 curved stamens. It is a common wildflower of open, moist, sandy sites from s. Que. southward; Jun.–Oct.

Maryland Meadow-beauty *(R. mariana)* is quite similar, but it has leaves that narrow toward their bases and pale pink flowers; May–Sep.

Meadow-beauty leaves can be eaten raw or cooked.

Goat's-rue
Tephrosia virginiana

Spurred Butterfly Pea
Centrosema virginianum

Virginia Meadow-beauty
Rhexia virginiana

Showy Evening-primrose   *Oenothera speciosa* Nutt.
Evening-primrose Family   Onagraceae

Evening-primroses (not related to true primroses of the primrose family) are perennials with a distinctive flower: 4 reflexed sepals and 4 petals attached at the end of a long calyx tube; the stigma has 4 branches that form a cross. The common name is derived from the several species that open toward evening (others, called "sundrops," open during the day). Sundrops and other evening-primroses are featured in the yellow section.

This species, also called White Evening-primrose, is our only species with white or pink flowers; its simple leaves are highly variable. It is a North American prairie plant that has become naturalized, especially along mountain roadsides, n. to PA; May, Jun.

Fireweed (Epilobe à Feuilles Étroites)   *Epilobium angustifolium* L.
Evening-primrose Family   Onagraceae

This tall (3–7 ft.), attractive plant, with its cluster of magenta flowers above narrowly lanceolate leaves, resembles Purple Loosestrife from a distance. Closer examination of a Fireweed flower, however, reveals 4 petals and an elongated pistil with a 4-branched stigma. Its name indicates its prevalence in recently burned sites, but it is also common in other open, disturbed places. It is found as far south as TN and NC but is more common northward; its northern range extends to the Gaspé and beyond; Jul.–Sep.

Other northern epilobiums with similar showy, magenta flowers are Hairy Willow-herb (*E. hirsutum*), with dentate leaves and notched petals, Jul.–Sep.; and River-beauty (*E. latifolium*), a shorter (1 ft.) plant with only a few large flowers (Gaspé, northward), Jun.–Sep.

Indian-pink   *Spigelia marilandica*
Logania Family   Loganiaceae

Also called Pink Root, this 1- to 2-ft.-tall perennial has 4–7 pairs of sessile, opposite, lanceolate leaves. Above them is the cluster of unusual tubular flowers, each 1–1½ in. long. It is found in rich woods and openings from MD, WV, and OH s. to GA, SC, and TN; Apr.–Jun.

An alternative name, "Wormgrass," refers to its widespread use by both American Indians and Western physicians as a worm purgative, especially for children. However, serious side effects, including death, may result from its use.

Showy Evening-primrose
Oenothera speciosa

Fireweed
Epilobium angustifolium

Indian-pink
Spigelia marilandica

Gay Wings   *Polygala paucifolia* Willd.
Milkwort Family   Polygalaceae

Gay Wings, a small (3–6 in. tall) plant with broad leaves and dainty, 1-in.-long flowers gives the initial appearance of an orchid. The 2 "wings" are sepals; the 3 petals are united into a tube tipped with a bushy fringe. Also called Flowering Wintergreen, it occurs (but infrequently) in rich, moist woods at low to middle elevations throughout the Apps; May–Jul.

Northern Wood-sorrel (Pain de Lievre) *Oxalis acetosella* L. *[O. montana]*
Wood-sorrel Family   Oxalidaceae

This distinctive plant forms colonies with flowers on separate stalks 2–4 in. long above the leaves. Note the delicate pink veins of the white petals. It is a common inhabitant of cool, moist, highly acidic soils of the spruce-fir forests of the s. Apps, but its range extends n. to Que., where it grows at low elevations; May–Aug.

Violet Wood-sorrel (*O. violacea*) is frequent in woodlands of the southern and middle Apps. It has violet petals and leaves that are purple underneath; Apr.–Jun.

The leaves of *Oxalis* species make a tart (due to oxalic acid) addition to salads and can be used to prepare a cold drink.

Wild Geranium   *Geranium maculatum* L.
Geranium Family   Geraniaceae

Not to be confused with the potted indoor "geraniums" (*Pelargonium* species) native to South Africa, *Geranium* species are plants of temperate Eurasia and N. Amer. They have dissected leaves and small to medium-size pink flowers.

Wild Geranium, our largest species, (1–2 ft. tall) has hairy 5-lobed leaves and pink to lavender 5-petaled flowers an inch across. It is common in rich woods and thickets from New Eng. southward; Apr.–Jun.

As the roots are rich in tannins, they have been used as a styptic and to treat cold sores.

Gay Wings
Polygala paucifolia

Northern Wood-sorrel
Oxalis acetosella

Wild Geranium
Geranium maculatum

Common Morning-glory
Morning-glory Family

Ipomoea purpurea (L.) Roth
Convolvulaceae

This twining vine, which may become 10 ft. long, has hairy stems and broadly cordate leaves. Funnel-shaped flowers (2–3 in. across) composed of 5 fused petals may be white, pink, purple, blue, or variegated. Introduced from tropical Amer., it is common in open sunny sites of the s. Apps and is less common northward; Jul.–Oct.

Other morning-glories include Red Morning-glory (*I. coccinea*), which has small scarlet flowers and cordate leaves, Jul.–Oct.; and Ivy-leaved Morning-glory, (*I. hederacea*), with white or rose flowers and leaves with 3 pointed lobes, Jul.–Sep.

Man Root (*I. pandurata*) is a native plant similar to *I. purpurea* but has white flowers with purple throats. Its huge (to 30 lbs.) roots have been used as a food source and to prepare medicines for a variety of ailments; Jun.–Sep.

False Dragonhead
Mint Family

Physotegia virginiana (L.) Benth.
Lamiaceae

This wiry upright perennial can grow to 4 ft. but is usually less than 3 ft. Leaves 3 to 5 in. long are lanceolate with dentate margins. Flowers (1 in. long) vary in color from light pink to deep rose; each has a spotted lower lip. It is common along riverbanks and other moist sites from Que. and New Bruns. southward; Jun.–Sep.

An alternative name, Obedient Plant, recognizes a unique feature: flowers, when repositioned on the stem, retain the new position.

Rose-pink
Gentian Family

Sabatia angularis (L.) Pursh
Gentianaceae

This highly branched, 1- to 3-ft.-tall biennial has thick, 4-angled stems. The opposite leaves are large and broadly lanceolate near the bottom of the plant, becoming smaller upward. The fragrant flowers, about 1 in. across, vary from the pink seen here to almost white; the greenish star in the middle of each is characteristic of *Sabatia* flowers. Rose-pink is fairly common in the Apps n. to NY, less so from there to s. New Eng. Its habitat includes open fields and woods; Jul.–Sep.

Also widespread but less common is the Slender Marsh-pink (*S. campanulata*), which has linear leaves and longer sepals; Jul., Aug.

Common Morning-glory
Ipomoea purpurea

False Dragonhead
Physotegia virginiana

Rose-pink
Sabatia angularis

Common Milkweed (Herbe à Coton)   *Asclepias syriaca* L.
Milkweed Family   Asclepiadaceae

In addition to their latex-like sap, milkweeds have small, intricate flowers, arranged in umbels. A typical flower has 5 reflexed petals above which are 5 erect hoods that form the corona (crown). Fruits are pods that enclose numerous plumed seeds.

Common Milkweed is a stout, 3- to 5-ft.-tall plant with large (6–8 in. long) ovate leaves. Flowers vary from white to dark pink. The pods are warty and pointed; Jun.–Aug.

It is a native species found throughout our area in open waste places. The young shoots and tender pods can be cooked as vegetables.

Swamp Milkweed   *Asclepias incarnata* L.
Milkweed Family   Asclepiadaceae

In comparison with Common Milkweed (*above*), this plant has narrower leaves and smaller flowers in somewhat flattened umbels. It is found in swamps and other wet sites throughout most of our area, especially at lower elevations; Jun.–Aug.

Wild Sweet-William   *Phlox maculata* L.
Phlox Family   Polemoniaceae

Phloxes are upright or trailing leafy perennials with opposite leaves and small but showy flowers that are usually pink or blue. Each flower includes 5 petals united into a corolla tube from which 5 separate petals extend at right angles.

This species is a 1½- to 3-ft.-tall plant with compact pink (or purple), nearly cylindrical flower clusters. The stem is purple-spotted. Also called Meadow Phlox, it is found along stream banks and in moist woods, as well as in meadows, from s. CT and NY s. to NC and TN; May–Sep.

These phloxes also have pinkish flowers, but they are arranged in less compact flower clusters: Garden Phlox (*P. paniculata*), which escapes from cultivation and has wider leaves with more conspicuous veins, Jul.–Sep.; Smooth Phlox (*P. glaberrima*), a smoother plant with leaves that are narrower and more pointed, Jun., Jul.

Downy Phlox   *Phlox pilosa* L.
Phlox Family   Polemoniaceae

The very narrow leaves of this 10- to 20-in.-tall perennial help to distinguish it from other phloxes. Note also the wedge-shaped petals, each about ¾ in. wide. "Downy" refers to the softly pubescent corolla tubes. It is infrequent in dry woods and open places from s. New Eng. s. to TN and SC; Apr.–Jun.

Common Milkweed
Asclepias syriaca

Swamp Milkweed
Asclepias incarnata

Wild Sweet-William
Phlox maculata

Downy Phlox
Phlox pilosa

Cuthbert's Turtlehead   *Chelone cuthbertii* Small
Figwort Family   Scrophulariaceae

Plants of this group have large, bilabiate (2-lipped) flowers that suggest the heads of turtles. In this species the leaves are sessile and have sharply serrate margins. Flowers are pink to red. It is a plant of bogs and moist upland forests of NC and VA; Jun.–Sep.

Other pink-flowered turtleheads of the s. Apps include *C. obliqua*, which has narrower leaves, Aug.–Oct.; and *C. lyonii*, with petioles up to 1 in. long, Aug.–Oct.

Balmony (*C. glabra*), a widespread species with white flowers tinged with pink, has been used to make a soothing ointment for piles, sore breasts, and ulcers; Aug.–Oct.

Twinflower (Linnee Boréale)   *Linnaea borealis* L. var. *americana*
Honeysuckle Family   Caprifoliaceae

This creeping evergreen plant has slender horizontal stems from which the 3- to 5-in.-tall stalks extend upward with their leaves and flowers. The ovate leaves are paired, as are the fragrant, nodding, 5-lobed, bell-like flowers. The only species of the genus, twinflowers are found in cool northern woods and mountains of e. N. Amer. from WV, OH, PA, and MD northward (also in Scandinavia).

The American variety of this dainty little plant differs only slightly from the European variety, which was a favorite of the great botanist Carolus Linnaeus (who named it for himself). Twinflower hasn't been seen in the s. Apps since it was reported by Albert Ruth in the mountains of Sevier County TN in 1892; Jun.–Aug.

Hollow Joe-Pye-weed   *Eupatorium fistulosum* Barratt
Aster Family   Asteraceae

Several tall *Eupatorium* species are called Joe-Pye-weed. (Joe Pye was a 19th-century New Englander who promoted the medicinal uses of plants, which he had learned from Indians.) They are 2–7 ft. tall, thick-stemmed plants with pink or purple flower heads in dense clusters. Leaves are toothed and whorled. They inhabit wet sites.

Hollow Joe-Pye-weed has rounded flower clusters and hollow, purple stems. It is found from s. ME and NY southward; Jul.–Sep.

Other pink or lavender Joe-Pye-weeds have solid stems and more flat-topped flower clusters: Spotted Joe-Pye-weed (*E. maculatum*), with purple-spotted stems, Jul.–Oct.; and Sweet Joe-Pye-weed (*E. purpureum*), with green stems, Jul.–Oct. Other *Eupatorium* species are featured in the white section.

Joe-Pye-weeds were used by Indians to treat a variety of urinary tract ailments, especially kidney stones.

Cuthbert's Turtlehead
Chelone cuthbertii

Twinflower
Linnaea borealis

Hollow Joe-Pye-weed
Eupatorium fistulosum

Great Rhododendron   *Rhododendron maximum* L.
Heath Family   Ericaceae

Heaths are shrubs or small trees with showy, urn-shaped flowers in clusters and alternate, simple leaves. As their root systems depend on soil fungi to aid in the absorption of water and nutrients, their distribution is often determined by the acidic soils necessary for the growth of fungi. Within the genus *Rhododendron* evergreen species are commonly called "rhododendrons" and deciduous species "azaleas."

Also called Great Laurel and Rosebay, Great Rhododendron is a common large shrub or small tree with 3- to 8-in.-long leathery leaves that taper to a point at both ends. The pink to rose purple flowers, each about 1 in. across, form spectacular clusters. It is found under hemlocks on moist slopes and streamsides, often forming dense tangles. Elevations are generally below 3,000 ft. but up to 5,000 ft. in the s. Apps. Its range extends from n. GA up the Apps into s. Que.; Jun., Jul.

Mountain Rosebay   *Rhododendron catawbiense* Michx.
Heath Family   Ericaceae

This common shrub, also called Catawba Rhododendron, is similar to Great Rhododendron (*above*), but it has leaves that are a little wider and less pointed at their tips. Flowers are somewhat larger, deeper in color, and arranged into larger clusters. It is a plant of middle to upper elevations (1,500–6,000 ft.) of the Apps, where it inhabits steep rocky slopes. It is often found in heath balds and is a component of both northern hardwood and spruce-fir forests. Its range extends from GA n. to VA and s. WV; May, Jun.

All parts of this and other *Rhododendron* species are poisonous if ingested; even honey made from their flowers may be toxic.

Piedmont Rhododendron   *Rhododendron minus* Michx. [*R. carolinianum*]
Heath Family   Ericaceae

As compared to the two "rhodies" above, this smaller but also evergreen shrub has smaller leaves (2–4 in. long) that are rusty underneath and smaller flowers in smaller clusters. It is found at both lower and higher elevations, especially on exposed ridges and balds, but is not common. It is known primarily in the mountains of NC, SC, GA, and TN; Apr.–Jun.

The three species featured here are probably the only evergreen rhododendrons native to the Appalachians. Deciduous species begin on the following page.

Great Rhododendron
Rhododendron maximum

Mountain Rosebay
Rhododendron catawbiense

Piedmont Rhododendron
Rhododendron minus

Pinxter Flower   *Rhododendron periclymenoides* (Michx.) Shinn.
[*R. nudiflorum*]
Heath Family   Ericaceae

Also called Pink Azalea, this small (to 8 ft. tall) shrub produces its white to pale pink flowers before or as the leaves open. Note the stamens, which are 2–3 times the length of the corolla tube. Leaves are 2–3½ in. long. It is found on cliffs and in woods or swamps n. to NY and New Eng; Mar.–May.

Swamp Azalea (*R. viscosum*) is a northern shrub with white flowers that open after the leaves; Jun., Jul.

Early Azalea   *Rhododendron prinophyllum* (Small) Millais
[*R. roseum* (Loisel.) Rehder]
Heath Family   Ericaceae

This beautiful azalea grows to 12 ft tall. Its fragrant flowers, which appear with the leaves, are pale to bright pink. Stamens are twice the length of the corolla tube. Early Azalea can be told from Pinxter Flower (*above*) because of its shorter corolla tube and its leaves, which are wooly underneath. In need of protection, it is infrequent in dry oak forests from ME s. to VA and KY; May, Jun.

Piedmont Azalea (*R. canescens*) is similar to Early Azalea but has narrow leaves; May.

Pink-shell Azalea   *Rhododendron vaseyi* Gray
Heath Family   Ericaceae

This tall (to 12 ft.) deciduous shrub produces its pink or rose flowers (note curved stamens) before the leaves unfold. Leaves, which are elliptical and taper to a sharp tip, distinguish this azalea from others. It is native only to bogs and spruce forests of NC but is cultivated in MA, where it has become naturalized; May, Jun.

Pinxter Flower
Rhododendron periclymenoides

Early Azalea
Rhododendron prinophyllum

Pink-shell Azalea
Rhododendron vaseyi

Rhodora (Rhododendron du Canada)   *Rhododendron canadensis*
                                                              (L.) Torr.
Heath Family   Ericaceae

Rhodora is a small (to 3 ft. tall) deciduous shrub with mostly upright branches. The flowers (½–1 in. across) open in spring before the narrow leaves unfold. It is a plant of acidic bogs, barrens, and rocky summits of the Gaspé, New Eng., NY, and n. PA; Mar.–Jul.

The scientific journal of the New England Botanical Society is named *Rhodora* in recognition of this attractive plant.

Lappland Rosebay (*R. lapponicum*) is a dwarf, prostrate plant with evergreen leaves and smaller, bell-shaped, magenta flowers. It is a circumboreal plant seen only on the highest peaks of our region: Katahdin, White Mountains, and the Adirondacks; Jun.

Sheep Laurel (Crevard de Moutons)   *Kalmia angustifolia* L. *[K. carolina]*
Heath Family   Ericaceae

A small erect shrub that grows to 3 ft. high, it has deep pink (rarely white) flowers clustered below the narrow, evergreen leaves at the top of the stem. Sheep Laurel is found at low to medium elevations of the n. Apps and at increasingly higher elevations southward. Another name, Lambkill, refers to its potential to poison livestock (also humans); May–Aug.

Pale Laurel (*K. polifolia*) is a similar plant easily distinguished from Sheep Laurel by its cluster of pink flowers at the top of the stem; May, Jun.

Mountain Laurel   *Kalmia latifolia* L.
Heath Family   Ericaceae

Also called Calico-bush and Ivy by mountain people, this is a large shrub or small (to 25 ft. tall) tree. The leaves are alternate but may appear whorled toward the ends of the branches. The white to pink saucer-shaped flowers are ¾ in. across. Mountain Laurel forms thickets at low to middle elevations (to 4,000 ft.), usually under hemlock or deciduous trees but also often in association with Mountain Rosebay Rhododendron (*above*) in mountain balds. It is found in the Apps n. to s. ME; May–Jul.

The unique *Kalmia* flower has spring-loaded stamens that deliver pollen. The 10 stamens are bent so as to fit into small grooves in the petals. When an insect touches a stamen, it springs forward releasing pollen onto the insect. Linnaeus named this genus in honor of his student Peter Kalm (1716–79), an early Appalachian explorer.

Rhodora
Rhododendron canadensis

Sheep Laurel
Kalmia angustifolia

Mountain Laurel
Kalmia latifolia

Deerberry
Heath Family

Vaccinium stamineum L.
Ericaceae

Deerberry is a highly branched, deciduous shrub that grows 2–4 ft. tall. Its thick leaves are 1–3 in. long. Numerous white or light pink flowers are less than ¼ in. long. The inedible berries that follow are greenish to pale purple. Also called Squawberry, it occupies dry woods from ME to GA; May, Jun.; fruits, Jul., Aug.

Cranberry (Atocas)
Heath Family

Vaccinium macrocarpon Aiton
Ericaceae

This is a creeping evergreen shrub with small, rounded or ovate leaves, ½ in. long, on highly branched, upright stems. The white to pink flowers resemble the head and neck of a crane. Tart red fruits are like those of cultivated cranberries, but smaller. A bog plant, it is common in the n. Apps, becoming less so southward but found as far s. as w. NC and e. TN; Jun.–Aug.; fruits, Sep.–Nov.

Indians, who called the berries "sassamanesh," used them for food and as a poultice on wounds. There are other cranberries native to Europe and Asia, but this is the species from which the cultivated cranberries of N. Amer. were developed.

Small Cranberry (*V. oxycoccus*) has similar flowers and fruits, but its leaves are smaller, have rolled edges, and are white beneath; May–Jul.; fruits, Aug.–Nov.

Mountain-cranberry (Pomme-de-Terre)
Heath Family

Vaccinium vitis-idaea L.
Ericaceae

Like other cranberries, Mountain-cranberry is a low (6 in. tall), creeping evergreen with slender stems. The ovate leaves, which average ½ in. long, have tiny black dots beneath, a feature that distinguishes the plant from other creeping shrubs. The species is a boreal plant of Europe, Asia, and N. Amer., with a range extension s. to the mountains of New Eng. It is locally abundant both in high alpine meadows and on rocky summits of lower mountains; Jun., Jul.; fruits, Aug.–winter.

Southern Mountain-cranberry (*V. erythrocarpon*) is a tall (to 8 ft.), upright shrub with larger, pointed leaves; it is a high-elevation plant of mountain forests from VA southward; May–Jul.; fruits, Sep., Oct.

Deerberry
Vaccinium stamineum

Cranberry
Vaccinium macrocarpon

Mountain-cranberry
Vaccinium vitis-idaea

Minnie-bush	*Menziesia pilosa* (Michx.) Juss.
Heath Family	Ericaceae

This low (to 6 ft. tall) shrub, like most heaths, has simple leaves clustered near the twig tips. Twigs are bristly, and the bark of older branches is "shreddy." The small urn-shaped flowers resemble those of some *Vaccinium* species *(above)*, but Minnie-bush can be distinguished from them and other heaths by a characteristic leaf feature: the midrib ends at the leaf tip in a small knoblike structure visible without a hand lens. Minnie-bush is common in spruce-fir forests and heath balds of VA, TN, NC, and GA; Jun., Jul.

Prairie Rose	*Rosa setigera* Michx.
Rose Family	Rosaceae

The rose family is a large temperate one that includes herbs, shrubs, and trees. Flowers typically have 5 green sepals, 5 rounded petals of various colors, and numerous stamens and pistils. The compound leaves are alternate and bear prominent stipules.

Roses (*Rosa* species) native to the Apps are shrubs with similar flowers (2–3 in. across and with 5 pink petals); thus they are identified primarily on the basis of their vegetative features. Prairie Rose is a climbing or sprawling rose distinguished by its 3 leaflets per leaf and single reflexed hooks on the stem. It inhabits open woods and waste places from s. NY southward; Jun., Jul.

Two other native roses are Swamp Rose *(R. palustris)*, with 5 or 7 leaflets and paired reflexed spines, Jun.–Aug.; and Pasture Rose *(R. carolina)*, also with 5 to 7 leaflets but slender straight spines, Jun., Jul.

Sweetbrier	*Rosa eglanteria* L.
Rose Family	Rosaceae

This alien rose bears more numerous but smaller flowers than those of our native roses. The fragrant leaves are divided into doubled-toothed (large and small teeth alternating), rounded leaflets. It grows to 6 ft. tall and is frequent in disturbed sites. Introduced from Europe, it is now found throughout e. N. Amer.; May–Jul.

Other alien roses seen in the Apps include Wrinkled Rose *(R. rugosa)*, a coarse shrub with wrinkled leaves, bristly stems, and deep pink (or white) flowers, Jun.–Sep.; and Multiflora Rose *(R. multiflora)*, with toothed stipules and clusters of small, white flowers; Jun.–Sep. The latter, especially, has become a problem because it often invades natural areas, smothering native flora.

A tea made from rose hips (fruits) is high in vitamin C.

Minnie-bush
Menziesia pilosa

Prairie Rose
Rosa setigera

Sweetbrier
Rosa eglanteria

Steeple-bush (Thé du Canada)
Rose Family

Spirea tomentosa L.
Rosaceae

This largely unbranched shrub has ovate, slightly toothed, alternate leaves that are wooly underneath and a spire-shaped panicle of pink flowers. Reaching a height of 2–4 ft., it grows along the edges of lakes and in other moist habitats. Also called Hardtack, it is frequent in the mountains from Que. to NC; Jul.–Sep.

Meadowsweet (*S. latifolia*) has similar toothed leaves and pale pink flowers in less compact, rounded panicles. It, too, is found throughout much of the Apps but is more common northward; Jun.–Aug. Narrowleaf Meadowsweet (*S. alba*) has narrower leaves and white flowers in rounded panicles; Jun.–Sep. Both of these small shrubs are found primarily in bogs and moist thickets.

Native Americans used a spirea leaf tea for the treatment of diarrhea and morning sickness. Various Eurasian spireas are cultivated as ornamental shrubs.

Queen-of-the-prairie
Rose Family

Filipendula rubra (Hill) Robinson
Rosaceae

This tall (to 6 ft.) plant resembles Steeple-bush (*above*) in general growth habit, but has much larger, sharply lobed, and deeply dissected leaves, and looser, fan-shaped flower clusters. More common in wet midwestern prairies, it also inhabits moist Appalachian habitats from NY s. to NC and KY; Jun., Jul.

Coral Honeysuckle
Honeysuckle Family

Lonicera sempervirens L.
Caprifoliaceae

The whorls of long (1–1½ in.) slender trumpets, coral with yellow lobes, are distinctive. Note also the paired leaves, said to be perfoliate, "pierced" by the stem. This smooth, twining vine is native to e. U.S. and often escapes from cultivation; it is found in woods and thickets from s. ME southward; Apr.–Sep.

Japanese Honeysuckle (*L. japonica*) is a weedy vine that "takes over" in disturbed areas. Its white flowers turn yellow as they mature (pink flowers occur occasionally). Long used medicinally in Asia, modern studies have shown that the flowers are, indeed, antibacterial and antiviral; Apr.–Jul.

Steeple-bush
Spirea tomentosa

Queen-of-the-prairie
Filipendula rubra

Coral Honeysuckle
Lonicera sempervirens

Cross Vine *Bignonia capreolata* L.
Bignonia Family Bignoniaceae

This woody vine has compound leaves, each with 2 ovate lateral leaflets and a terminal tendril. Note the large (3 in. long), orange to coral, trumpet-like flowers with yellow throats. Primarily a plant of the s. Apps, it is found in moist woods and swamps n. to s. OH and PA. Named for the X visible in a cross-section of its stem, it often climbs to the top of tall trees; Apr.–Jun.

Another vine of this family with large trumpetlike flowers is Trumpet-creeper *(Campsis radicans)*. It has leaves with 7–11 dentate leaflets and orange flowers; Jul.–Sep.

Silk-tree *Albizia julibrissin* Durazz
Mimosa Family Mimosaceae

The pink, plumelike flowers with their long, numerous stamens make this Asian tree quite distinct. Growing to 50 ft. tall, it has large (1 ft. or longer), doubly compound leaves. Both the leaves and fruits *(lower center)* are suggestive of the alliance of this family with the pea family; Jun.–Aug.

Although often called "Mimosa," Silk-tree is distinct from the small, native, shrubby Sensitive Briar *(Mimosa quadrivalvis)* of this family. It also has rounded, pink flowers, similar (but smaller) leaves, and elongated seed pods. Its common name is due to the leaves, which draw up quickly when touched; May–Sep.

Redbud *Cercis canadensis* L.
Caesalpinia Family Caesalpiniaceae

Before the leaves of this small tree appear, red buds open to form pink flowers. The leaves are simple, cordate, and 3–5 in. long. Redbud is a common understory tree, especially in low-elevation oak-hickory forests. Its range extends n. to s. New Eng.; despite *canadensis* as a part of its name, it does not occur naturally in Canada; Mar.–May.

The flowers give color and a pleasant tartness to salads. The tender young pods can be sauteed in butter.

Cross Vine
Bignonia capreolata

Silk-tree
Albizia julibrissin

Redbud
Cercis canadensis

Blue/Purple

Common Dayflower   *Commelina communis* L.
Spiderwort Family   Commelinaceae

Unlike spiderworts *(below)*, which have 3 petals of equal size, dayflowers have 2 prominent ones and a third reduced one. Each flower lasts only a day. This common sprawling Asian species is frequent in moist or shaded waste places and as a garden-weed from NH southward; May–Oct.

The somewhat less common Erect Dayflower (*C. erecta*) is a similar but more erect native plant; Jun.–Oct.

Dayflowers may be eaten as potherbs.

Virginia Spiderwort   *Tradescantia virginiana* L.
Spiderwort Family   Commelinaceae

Spiderworts are perennials that produce terminal clusters of flowers (1 in. across) with 3 identical blue, purple, or rose petals.

Virginia Spiderwort, also called Widow's-tears, has round, violet petals and yellow anthers. Note the hairy buds (*lower right*) and flower stalks. The linear leaves are 12–15 in. long and keeled. This showy plant grows in moist sites from ME to PA southward; Apr., May.

Smooth Spiderwort (*T. ohiensis*) has smooth buds and flower stalks, Apr.–Jul; Zigzag Spiderwort (*T. subaspera*) has zigzag stems and wider leaves, May–Jul. Both species are principally midwestern spiderworts but are seen along the w. edge of the middle and s. Apps.

Spiderworts can be used as indicators of nuclear radiation; filament hairs change from blue to pink in response to low-level dosages. They are also cultivated as garden ornamentals (named for John Tradescant, gardener to England's Charles I).

Pickerelweed   *Pontederia cordata* L.
Pickerelweed Family   Pontederiaceae

Pickerelweed, with its cordate leaves on long stalks and dense spikes of purple flowers, is quite showy. It grows rooted on the bottoms of shallow ponds and along the edges of lakes and streams, a habitat where pickerel (fish) swim and lay their eggs; thus Pickerelweed. It is found sporadically throughout our area; Jun.–Nov.

Found on separate plants are 3 kinds of flowers that differ in the lengths of their styles and stamens. This arrangement, known as tristyly, promotes cross-pollination.

Common Dayflower
Commelina communis

Virginia Spiderwort
Tradescantia virginiana

Pickerelweed
Pontederia cordata

Wild Hyacinth   *Camassia scilloides* (Raf.) Cory
Lily Family   Liliaceae

Wild Hyacinth has keeled, linear leaves at the base of 1- to 2-ft. racemes. These showy plants often occur in large numbers on basic soils of meadows and open deciduous woods. Although more common west of the mountains, they occur infrequently in the s. Apps n. to sw. PA; Apr.–Jun.

Camassia is derived from the Indian "quamash." The onionlike bulbs can be cooked and eaten (but are gummy).

Grape-hyacinth *(Muscari botryoides)* is a common garden ornamental that persists around old homesites. It is a much smaller plant; its numerous deep purple flowers are tiny balls; Apr., May.

Large Purple Fringed Orchid   *Habenaria psycodes* (L.) Sprengel var.
                      grandiflora Bigelow A. Gray *[Platanthera grandiflora]*
Orchid Family   Orchidaceae

Habenaria orchids produce small, often fringed flowers arranged in racemes or spikes. In this species the lower tips of the three petals are 3-parted and deeply fringed. The flower clusters are more than 2 in. thick and are borne on stalks 3–5 ft. tall. This orchid occurs sporadically at high elevations in meadows and seeps throughout the Apps, including the Adirondacks; Jun.–Aug.

Small Purple Fringed Orchid (*H.p.* var. *psycodes*) is not as tall (2–3 ft.) and has smaller flower clusters (less than 2 in. across); the two varieties sometimes hybridize; Jun.–Aug.

Blue-eyed Grass   *Sisyrinchium angustifolium* Mill.
Iris Family   Iridaceae

Blue-eyed grasses, of which there are six Appalachian species, are small herbs with narrow, grasslike leaves and tiny white to blue or violet starlike flowers on wiry stems. Each of the 3 sepals and 3 petals are alike, bearing a hairlike point that extends from the tip.

This species has light blue flowers atop flat stems, 6–12 in. tall and narrow (¼ in. wide). It is commonly found on moist, sandy soils, especially in open areas from MA southward; Mar.–Jun.

White Blue-eyed Grass *(S. albidum)* has white flowers; Mar.–May.

Wild Hyacinth
Camassia scilloides

Large Purple Fringed Orchid
Habenaria psycodes

Blue-eyed Grass
Sisyrinchium angustifolium

Northern Blue Flag (Clajeux) *Iris versicolor* L.
Iris Family Iridaceae

Irises are named for the Greek goddess of the rainbow, indicating the wide color variations observed in these plants. Iris flowers appear to have 9 petals; in reality, the outer 3 parts are sepals (called falls), the next 3 are erect petals (standards), and the inner 3 are colorful branches of the stigma. The prominent markings on the falls help direct insects or birds to the nectaries, where they make contact with the stamens and stigmas. The Iris is the state (cultivated) flower of TN and the floral emblem (fleur-de-lis) of Que. and France.

Northern Blue Flag, a 2- to 3-ft.-tall plant, occurs in wet meadows and lake shores in the mountains from Que. to VA; May–Aug.

The roots of this plant have been widely used medicinally both by American Indians and in homeopathy, but they can be poisonous.

Slender Blue Flag (*I. prismatica*) is very similar in coloration and markings but has narrower sepals, petals, and leaves; May, Jun.

Dwarf Crested Iris *Iris cristata* Ait.
Iris Family Iridaceae

This is a small, low-growing iris; leaves are only 6–9 in. long, and flowers are only 2½ in. across. Note the "bearded" sepals. It often occurs as extensive colonies in moist woods along bluffs and stream banks at low elevations from WV southward; Apr., May.

Native Americans made from the roots an ointment that was used to treat skin cancer.

Upland Dwarf Iris *Iris verna* L. var. *smalliana* Fern. ex M. E. Edwards
Iris Family Iridaceae

This fragrant iris can be distinguished from *I. cristata* (*above*) by its uncrested falls, which are marked but lack beards. Its leaves also are narrower (½ in. or less). The species grows more commonly along the Coastal Plain east of the mountains; this variety, characterized by somewhat wider leaves, grows in dry woods and openings from PA and WV southward; Apr., May.

Northern Blue Flag
Iris versicolor

Dwarf Crested Iris
Iris cristata

Upland Dwarf Iris
Iris verna

Common Blue Violet (Violette)   *Viola sororia* Willd. *[V. papilionacea]*
Violet Family   Violaceae

Violets, with their small, 5-petaled flowers (2 upper, 2 lateral, and 1 terminal) and simple leaves, are among the most recognizable groups of wildflowers. Identification to species, however, is often difficult because of variations within and hybridization between species. Features to consider in identification include leaf shape, flower color and markings, and whether the violet is "stemmed" (leaves, flowers on same stem) or "stemless" (on separate stems). In addition to some 40 species native to the Apps, there are also several European ones you might encounter.

This stemless violet is highly variable. A "typical" plant of the species, such as seen here, is 4–9 in. tall and has cordate leaves; both lateral petals are bearded, and the terminal petal is slightly longer and unbearded. It is common in meadows and moist woods (also as a weed in lawns) from s. Que. southward; Mar.–Jun.

Plants of this species that have gray petals with blue veins are called Confederate Violets.

Birdfoot Violet   *Viola pedata* L.
Violet Family   Violaceae

Also a highly variable, stemless violet, this low-growing (6 in.) but spreading plant has deeply dissected leaves that resemble a bird's foot. Its flowers are large (1–1¼ in. across), beardless, and in one variety, bicolored (purple/lavender). It is frequent in sunny, often dry, sites from s. New Eng. southward; Apr.–Jun.

The leaves of Birdfoot Violet were used by Indians to prepare an expectorant and for certain lung complaints; during the 19th century, it was used in a similar way by European physicians.

The less common Coast Violet (*V. brittonia*) also has "birdfoot" leaves, but its petals are reddish violet with white throats, and the 3 lower petals are markedly bearded; Apr.–Jun.

Wild-pansy   *Viola rafinesquii* Greene *[V. kitaibelliana* var. *rafinesquii]*
Violet Family   Violaceae

This 3- to 6-in.-tall stemmed violet bears spoon-shaped leaves with leaflike, divided stipules. The small (½ in. across), pansylike flowers have petals that vary from white to light blue with blue veins. It is a common, weedy alien of open, disturbed areas from NY southward; Apr.–Sep.

Called Field-pansy, *V. arvensis* is a coarser plant with hairy stems and leaves. Petals are yellow with purple veins. It occurs in cultivated fields throughout our area; Apr.–Sep.

Common Blue Violet
Viola sororia

Birdfoot Violet
Viola pedata

Wild-pansy
Viola rafinesquii

Marsh Blue Violet   *Viola cucullata* Aiton
Violet Family   Violaceae

This plant is similar to the Common Blue Violet (*above*), but it can be distinguished by its young leaves that are rolled inward, longer flower stalks that hold the flowers well above the leaves, and petals that are darker toward the center, with the terminal petal shorter and bearded. Also called Bog Violet, it grows in bogs and along stream banks throughout the Apps; toward the south, it occurs at increasingly higher elevations; Apr.–Jun.

Several violets, especially blue-flowered ones, are edible. Tender young leaves can be added to a salad or cooked as greens; dried ones can be used to make a tea. The flowers are sometimes candied and eaten as a delicacy.

Long-spurred Violet   *Viola rostrata* Pursh
Violet Family   Violaceae

This stemmed violet, which grows 4–8 in. tall, has lavender to purple flowers held well above the leaves. Each flower has veined, beardless petals and a long ($\frac{1}{2}$ in.) slender spur. It occurs in rich, often calcareous, forests from New Eng. southward; Apr.–Jun.

The Dog Violet (*V. conspersa*) is similar but has shorter ($\frac{1}{4}$ in. or less) spurs and bearded lateral petals; Apr.–Jun.

Passion Flower   *Passiflora incarnata* L.
Passion Flower Family   Passifloraceae

A member of a principally tropical family, this sprawling or climbing vine is one of the few species of the family in temperate N. Amer. The flowers (to 2 in. across) are not easily described but are so distinctive as to be easily recognized; the corona (an outgrowth between the stamens and petals) is divided into numerous fringelike purple/white segments. Passion Flower grows in fencerows and other sunny, disturbed areas, principally at low elevations, n. to PA; May–Sep.; fruits, Jul.–Oct.

The names Maypop and Wild Apricot refer to the ripe fruits, which are delicious and can be eaten or made into a cold drink. Native Americans poulticed the roots for wounds and inflammations and used the tops to make a tea to treat insomnia, neuralgia, and epilepsy. Modern research confirms its efficacy for such conditions. Passion Flower is the state wildflower of TN.

The less common Yellow Passion Flower (*P. lutea*) is a vine with similar leaves but much smaller, greenish yellow flowers; May–Sep.; fruits, Jul.–Oct.

Marsh Blue Violet
Viola cucullata

Long-spurred Violet
Viola rostrata

Passion Flower
Passiflora incarnata

Dwarf Larkspur *Delphinium tricorne* Michx.
Buttercup Family Ranunculaceae

Larkspurs are so named because one of the petal-like sepals forms a "spur"; petals are absent. Leaves are dissected somewhat like those of buttercups. This 1- to 2-ft.-tall plant produces flowers that vary from purple to pink to white. It is found in calcareous soils of woodlands from PA southward; Apr.–Jun.

Tall Larkspur (*D. exaltum*) is a taller plant (2–6 ft.) that blooms later. It has blue or white flowers and similar but larger leaves; Jul., Aug.

Delphinium species, like many others of this family, are poisonous if eaten (grazing cattle are often affected).

Purple Loosestrife (Bouquet Violet) *Lythrum salicaria* L.
Loosestrife Family Lythraceae

This 2- to 4-ft.-tall perennial has lanceolate leaves beneath a showy spike of magenta flowers. Each flower has 6 petals (unusual for dicots). A European plant imported as an ornamental, it has spread aggressively, often at the expense of native vegetation. More common in wet meadows and pond and lake edges of the n. Apps, it is spreading southward. Recently it has been spotted, to the consternation of environmentalists, in middle TN; Jun.–Sep

A tea made from the plant has been shown to be antibacterial, perhaps accounting for its use in European folk medicine to treat wounds and as an astringent.

Narrow-leaved Vervain *Verbena simplex* Lehmann
Vervain Family Verbenaceae

Vervains are small weedy plants of open waste places. They have simple leaves below a spike of tiny, 5-petaled flowers that are usually pink or purple.

Narrow-leaved Vervain has lavender or purple flowers and toothed, narrowly lanceolate leaves. It is found from s. Que. southward; May–Sep.

Vervains that have flowers similar to those of *V. simplex* include Hoary Vervain (*V. stricta*), with toothed ovate leaves, Jun.–Sep.; and Blue Vervain (*V. hastata*), with leaves similar to those of *V. simplex* but wider and with more prominent teeth, Jun.–Oct. Two species with deeply dissected leaves include the weedy European Vervain (*V. officinalis*), Jun.–Sep.; and our native Bank Vervain (*V. riparia*), Jun., Jul. White Vervain (*V. urticifolia*) is distinctive because of its flower color; Jun.–Oct.

The medicinal use of Blue Vervain by American Indians and of European Vervain by Europeans and Asians suggests the medical potential of other verbenas.

Dwarf Larkspur
Delphinium tricorne

Purple Loosestrife
Lythrum salicaria

Narrow-leaved Vervain
Verbena simplex

Indigo Bush   *Amorpha fruticosa* L.
Pea Family   Fabaceae

Plants of the pea family are generally recognized by their compound leaves and flowers that suggest those of sweet peas followed by beanlike pods. Included are trees, shrubs, vines, and herbs.

Indigo Bush, also called False Indigo, is an erect shrub with deciduous leaves that are fragrant when crushed. The numerous (13–25) leaflets have small dots underneath. The purple flowers are unusual in that each has only one petal. Fruits are less than ½ in. long and warty. Growing to 15 ft. tall, it forms thickets along stream banks throughout the Apps; May, Jun.

Mountain-indigo (*A. glabra*), a southern Appalachian endemic, occupies drier sites. It can be distinguished from *A. fruticosa* by its flowers, each of which has one or more prominent calyx lobes; May, Jun.

Kudzu   *Pueraria lobata* (Willd.) Ohwi
Pea Family   Fabaceae

This noxious weedy vine is a native of Asia. Each leaf has 3 wide 3-lobed leaflets. Racemes of sweet-smelling violet flowers, often hidden behind the leaves, are only 3–4 in. long. It is common in the se. U.S., where it was often planted for erosion control. Although its range is extending northward to and possibly beyond PA, it does not usually flower n. of VA; Jul.–Sep.

Kudzu, which smothers crops and native vegetation in the U.S., is widely used in Chinese medicine.

Soapwort Gentian   *Gentiana saponaria* L.
Gentian Family   Gentianaceae

Gentians are erect or sprawling leafy herbs that flower in late summer or early fall. The opposite leaves are simple. Flowers have 4 or 5 green sepals and 4 or 5 blue petals fused to form a bell-shaped corolla.

In Soapwort Gentian, several blue violet flowers are clustered at the nodes, which also bear whorls of lanceolate leaves. Its leaves resemble those of Soapwort, hence the name. It inhabits moist glades and wet places from NY and WV southward; Aug.–Oct.

Each of these gentians has one or more features that can be used to differentiate it from the above: Fringed Gentian (*G. crinata*) has 4 delicately fringed petals that flare from the corolla, Sep.–Oct.; Narrow-leaved Gentian (*G. linearis*) has very narrow leaves, Aug., Sep.; and Stiff Gentian (*G. quinquefolia*) has wider leaves and more tubular, lilac flowers, Aug.–Oct.

Indigo Bush
Amorpha fruticosa

Kudzu
Pueraria lobata

Soapwort Gentian
Gentiana saponaria

Forest Phlox *Phlox divaricata* L.
Phlox Family Polemoniaceae

This hairy plant is less than 2 ft. tall and has small, opposite, lanceolate leaves. The flowers, 1 in. across, vary from light blue to pale violet. Forest Phlox inhabits open woods and rocky slopes from New Eng. s. to NC and TN; Apr.–Jun.

Shale-barren Phlox (*P. buckleyi*) has bright blue flowers; May, Jun.

Jacob's-ladder *Polemonium reptans* L.
Phlox Family Polemoniaceae

Upright or partially reclining, Jacob's-ladder grows to about 1 ft. tall. Note the compound leaves with pointed leaflets and the small (½ in. across) flowers with white stamens. Also called Greek Valerian, it grows in moist mountain sites, especially along streams. It occurs from the Adirondacks southward but is more common w. of the Apps; Apr.–Jun.

Canadian Waterleaf *Hydrophyllum canadense* L.
Waterleaf Family Hydrophyllaceae

Members of this genus have white to light blue flowers, ½ in. across, and prominent stamens that are longer than the corolla. Identification to species is based primarily on leaf shape.

Canadian Waterleaf has flowers held well above its maplelike leaves. It is found in rich, moist woods from New Eng. southward; May, Jun.

Appendaged Waterleaf (*H. appendiculata*) has leaves with 5 more-distinct lobes; it is found along the w. edge of the Apps from PA to TN; May, Jun. More widespread *Hydrophyllum* species, both with compound leaves, include Virginia Waterleaf (*H. virginianum*), whose leaves have 3–7 leaflets, May, Jun.; and Large-leaved Waterleaf (*H. macrophyllum*), which bears 9–13 leaflets, May, Jun.

Purple Phacelia *Phacelia bipinnatifida* Michx.
Waterleaf Family Hydrophyllaceae

Related to the waterleaves (*above*), phacelias are similar but have narrower leaves and flowers in less dense clusters. Purple Phacelia is a 2-ft.-tall, upright biennial. Note the orange-tipped stamens that extend beyond the blue flowers, which are ½ in. across, and the segmented, coarsely toothed leaves. It is a plant of rich shaded woods from VA, WV, and OH southward; Apr., May.

Appalachian Phacelia (*P. dubia*) is a smaller plant with smaller blue, non-fringed petals and much smaller leaves; Apr., May. Miami-mist (*P. purshii*) has fringed, blue petals; Apr.–Jun.

Forest Phlox
Phlox divaricata

Jacob's-ladder
Polemonium reptans

Canadian Waterleaf
Hydrophyllum canadense

Purple Phacelia
Phacelia bipinnatifida

Lyre-leaved Sage *Salvia lyrata* L.
Mint Family Lamiaceae

Members of this family are distinctive because of their "square stems" (4-sided as seen in cross-section). Most have opposite leaves and are aromatic. Bilabiate (2-lipped) flowers, in clusters, have upper and lower lips that guard the opening into the corolla tube; inside are 2 long stamens and 2 short ones.

Lyre-leaved Sage is named for its rosette of basal leaves, each of which is shaped somewhat like the stringed instrument. Flower stalks average a foot in height; flowers are 1 in. long. The slightly aromatic plant is found along roadsides and waste places from s. New Eng. and NY southward; Apr.–Jun.

Also called Cancerweed, *S. lyrata* has been used in folk medicine to treat cancer and warts.

Nettle-leaved Sage (*S. urticifolia*) has dentate but undissected leaves attached along the flower stalk and medium blue flowers. It is found from PA and OH southward; May, Jun.

Meehania *Meehania cordata* (Nutt.) Britt.
Mint Family Lamiaceae

One who locates this handsome but uncommon 6-in.-tall perennial herb is fortunate indeed. The blue flowers, each 1 in. long, are borne in one-sided spikes above the cordate leaves. It forms large colonies on basic soils of rich woods from w. PA to TN and NC; Jun.

The genus, which includes only this species in Appalachia, was named for Thomas Meehan (1826–1901), a Philadelphia botanist.

Wild Bergamot *Monarda fistulosa* L.
Mint Family Lamiaceae

Flowers of this 2- to 3-ft.-tall plant vary from pink to lilac; bracts (underneath) are often lilac also. Probably our most common *Monarda* species, it grows in dry forest openings and edges, especially in calcareous regions, from s. Que. southward; Jul., Aug.

This showy perennial is often cultivated. An oil extracted from its leaves has been used for the treatment of respiratory ailments; an aromatic tea can be made from its leaves.

Basal Bee Balm (*M. clinopodia*) is very similar but has white to pink flowers and white bracts; Jun.–Sep.

Lyre-leaved Sage
Salvia lyrata

Meehania
Meehania cordata

Wild Bergamot
Monarda fistulosa

Heal-all (Herbe au Charpentier)   *Prunella vulgaris* L.
Mint Family   Lamiaceae

Note the delicate flowers with hooded, purple, upper lobes and fringed, white, lower ones interspersed among bracts in the oblong flower head. The plant is low (½–1 ft. tall) and often creeps. Leaves are lanceolate. Flower color varies from nearly white to deep purple. Native to Europe, it is now well established in disturbed soils throughout most of the temperate N. Hemisphere; May–Sep.

Clingman's Hedge-nettle   *Stachys clingmanii* Small
Mint Family   Lamiaceae

Hedge-nettles have several whorls of flowers on the stalk that also bears the leaves. The name comes from the resemblance of the leaves to those of nettle (but they lack stinging hairs).

Note the spotted petals, reddish stems, and dentate leaves of Clingman's Hedge-nettle. This rare plant, which grows 2–3 ft. tall, honors Thomas L. Clingman (1812–97), for whom Clingman's Dome of the Great Smoky Mountains is also named. It occupies high mountain peaks of the s. Apps, where it becomes locally abundant by spreading from stolons (horizontal stems); Jun.–Aug.

Showy Skullcap   *Scutellaria serrata* Andr.
Mint Family   Lamiaceae

Skullcaps are so named because of the small protuberance on the upper lip of the calyx, suggesting a cap or helmet. Showy Skullcap, 1–2 ft. tall, has long (1 in.) purple flowers in a terminal raceme. Note the serrate margins of the ovate leaves. It occupies rich woods from se. NY to SC and TN; May, Jun.

Of the dozen or so additional skullcaps, the most distinctive is Mad-dog Skullcap (*S. laterifolia*). It is recognized by its one-sided racemes, which arise from leaf nodes; Jul.–Sep.

Forget-me-not   *Myosotis scorpioides* L.
Borage Family   Boraginaceae

Forget-me-nots are easily recognized by their small, 5-petaled blue flowers on 2 branches. This European native, with flowers ⅓ in. across, has become naturalized along stream banks and in other wet places throughout the Apps. It is also cultivated as a garden perennial; May–Sep.

The native Smaller Forget-me-not (*M. laxa*) is similar but has smaller flowers (⅕ in. across); May–Oct.

Heal-all
Prunella vulgaris

Clingman's Hedge-nettle
Stachys clingmanii

Showy Skullcap
Scutellaria serrata

Forget-me-not
Myosotis scorpioides

Virginia Bluebells   *Mertensia virginica* (L.) Pers.
Borage Family   Boraginaceae

Members of the borage family are noted for their one-sided, rolled-up flower coils that unfurl as the flowers mature. Each flower has 5 sepals, 5 petals, and 5 stamens. The simple leaves are alternate.

Virginia Bluebells (not related to the bluebells of the Campanulaceae) is a distinctive 1- to 2-ft.-tall plant with pink buds that open into blue, bell-like flowers. It often forms large colonies along stream banks and on other alluvial soils from NY s. to KY, TN, and VA; Mar.–Jun.

Wild Comfrey   *Cynoglossum virginianum* L.
Borage Family   Boraginaceae

Flower clusters that suggest forget-me-nots combined with large, clasping leaves identify this 1- to 2-ft.-tall plant. It is found in various types of deciduous forests from NY and s. New Eng. southward; May, Jun.

Cherokees used a root tea for a great variety of ailments including cancer. Physicians of the 19th century used it as a substitute for Comfrey.

Northern Wild Comfrey (*C. boreale*) is a smaller, more slender plant; it has a range north of that of Wild Comfrey; May, Jun.

Comfrey (*Symphytum officinale*) is a European plant similar to Wild Comfrey. A distinguishing feature is its leaves, which taper toward the stem (rather than clasping it). Flower clusters are also more coiled. Its ancient use to treat wounds, bruises, etc. is substantiated by the presence of allantoin, a healing compound; May–Sep.

Blue Dogbane   *Amsonia tabernaemontana* Walt.
Dogbane Family   Apocynaceae

Also called Blue-star, this and other *Amsonia* species are herbs with alternate leaves and sky-blue, starlike flowers. Blue Dogbane, which grows to 3 ft. tall, has flowers ½ in. across. Its sharply lanceolate leaves are smooth and crowded on the stem. It grows in moist woods and along riverbanks of the s. Apps; Apr., May.

Virginia Bluebells
Mertensia virginica

Wild Comfrey
Cynoglossum virginianum

Blue Dogbane
Amsonia tabernaemontana

Gray Beardtongue   *Penstemon canescens* Britt.
Figwort Family   Scrophulariaceae

Members of this family have paired leaves and flowers in which the 5 (or 4) petals are united into a bilabiate (2-lipped) corolla tube suggestive of a mint (Lamiaceae). The stems are round, however, and the shoots lack a minty fragrance. The roots of many species are parasitic on the roots of other plants.

Beardtongues are erect herbs, often with sessile leaves. Flowers have 5 stamens; the longest one, prominently bearded, extends beyond the corolla tube; thus "beardtongue."

Note the open throats of the rose violet flowers of Gray Beardtongue. The 1- to 3-ft.-tall plants are covered with fine gray, hairs. It thrives in woods from PA southward; May, Jun.

Other beardtongues with brightly colored flowers include Hairy Beardtongue (*P. hirsutus*), which is very similar but has a hairy stem and flowers with their throats nearly closed, May–Jul.; and Smooth Penstemon (*P. laevigatus*), a less hairy plant with purple flowers, May, Jun.

Blue-eyed Mary   *Collinsia verna* Nutt.
Figwort Family   Scrophulariaceae

Each bicolored flower has 5 lobes: 2 upper white ones and 2 lower blue ones, with the fifth concealed between. Note the toothed, sessile, lanceolate leaves in whorls. Not common, but sometimes growing in great profusion along shaded stream banks, its range extends from NY s. to VA and (barely) TN; Apr.–Jun.

Cordillean Blue-eyed Mary (*C. parvifolia*) of w. New Eng. has blue/yellow flowers; May–Jul.

Monkey-flower   *Mimulus ringens* Ait.
Snapdragon Family   Scrophulariaceae

The shape of the flowers accounts for the unusual name. In this species, violet flowers are borne on long stalks; leaves are sessile; stems are 4-angled. It is a wetland plant found throughout our area; Jun.–Sep.

Another monkey-flower, *M. alatus*, also a 1- to 3-ft.-tall plant, has white, pink, or violet flowers on short stalks and stalked leaves; Jun.–Sep.

Gray Beardtongue
Penstemon canescens

Blue-eyed Mary
Collinsia verna

Monkey-flower
Mimulus ringens

Smooth Ruellia
Acanth Family

Ruellia strepens L.
Acanthaceae

Ruellias are herbs with opposite leaves and blue, trumpetlike flowers. From each slender corolla tube extend 5 flaring lobes. At the base of each flower is a pair of small leaves. Smooth Ruellia is a 1- to 3-ft.-tall plant that grows in semishaded places from OH and WV southward; May–Jul.

Hairy Ruellia (*R. caroliniensis*) has hairy leaves and stems; corolla tubes are somewhat longer and stalkless; Jun.–Sep.

Water-willow
Acanth Family

Justicia americana (L.) Vahl
Acanthaceae

This aquatic and wetland plant has slender, opposite leaves with clusters of bicolored (lilac/white) flowers. It often forms large colonies in shallow water along the edges of lakes and slow-moving streams. Its range extends from New Eng. southward; Jun.–Oct.

Tall Bellflower
Bluebell Family

Campanula americana L.
Campanulaceae

Bellflowers are annual herbs with alternate, simple, lanceolate leaves with toothed margins and often bell-like (sometimes not) flowers, usually blue. Tall Bellflower is a tall (2–6 ft.) plant with both leaves and flowers along its stalk. Flowers (1 in. across) are not bell-shaped but rather composed of 5 widely spread petals; note also the long, curved style. It commonly occupies shaded, moist sites from NY southward; Jun.–Aug.

Both Harebell (*C. rotundifolia*) and Southern Harebell (*C. divaricata*) have tiny, dark blue, bell-like flowers that dangle from the ends of branches. In the former, leaves (except for rounded basal leaves) are linear; in the latter, leaves are sharply lanceolate with a few large teeth; Jun.–Sep. (both).

Great Lobelia
Bluebell Family

Lobelia siphilitica L.
Campanulaceae

Related to bellflowers (*above*), lobelias are also herbs with simple, alternate leaves. Their tubular flowers have 2 narrower lobes above 3 wider ones. Great Lobelia, 1–3 ft. tall, is the largest of the group. The large (1 in. long) blue flowers with white stripes on the corolla tubes make identification certain. It is found in wet sites from ME southward; Aug.–Oct.

American Indians used a root tea made from this plant for the (ineffective) treatment of syphilis.

Other lobelias also with blue flowers: Indian Tobacco (*L. inflata*) has smaller, swollen-based flowers in leaf axils, Jul.–Oct.; Downy Lobelia (*L. puberula*) is a hairy plant with flowers in a 1-sided spike, Aug.–Oct. Pale-spike Lobelia (*L. spicata*) has small, white to light blue flowers in a spike; May–Jul.

Smooth Ruellia
Ruellia strepens

Water-willow
Justicia americana

Tall Bellflower
Campanula americana

Great Lobelia
Lobelia siphilitica

Large Houstonia   *Hedyotis purpurea* (L.) T. & G. *[Houstonia purpurea]*
Madder Family   Rubiaceae

Its size (½–1½ ft. tall) distinguishes this from the smaller *Hedyotis* species. It has flowers with 4 petals arranged like a cross and paired, ovate, sessile leaves below the terminal clusters of pale violet, tubular flowers. It is a plant of dry woods, often on sandy or rocky soils from VA and WV southward; Apr.–Jul.

Mountain Bluet (*H. montana*), a rare plant of high-elevation rocky outcrops of TN and NC, has deep purple flowers; Jun.–Aug. Long-leaved Bluet (*H. longifolia*) has narrower leaves and smaller clusters of white or lavender flowers; May–Jul.

Bluet   *Hedyotis caerulea* (L.) Hook. *[Houstonia caerulea]*
Madder Family   Rubiaceae

Bluets are small plants with tiny, opposite leaves and blue 4-petaled flowers. They grow in great profusion in open, moist, or grassy places. Note the bluish white petals with their yellow centers; flowers of this species are borne on short, erect stalks. It occurs in meadows throughout the Apps; Mar.–Jul.

High-elevation bluets include Thyme-leaved Bluet (*H. serpyllifolia*), a very low-growing plant with prostrate stems, Apr.–Jul.; and *H. pusilla* with red-centered flowers; Apr.–Aug.

Princess Tree   *Paulownia tomentosa* (Thumb.) Steudel
Bignonia Family   Bignoniaceae

This fast-growing deciduous tree, with large (5–10 in. long) cordate leaves, can reach a height of 100 ft. The fragrant tubular flowers, which appear before the leaves, are violet and have 5 rounded lobes. Fruits are hard, dry pods that contain winged seeds. A Chinese tree that was accidentally introduced into N. Amer. during the 1840s, it is now naturalized in open places from s. NY southward; Apr., May.

Wild Teasel (Cardere) *Dipsacus sylvestris* Hudson *[D. fullonum]*
Teasel Family   Dipsacaceae

Unlike the florets of the aster family, those of teasel have nonfused stamens. Note the numerous spiny bracts that project beyond the white to lavender florets. Not visible are the upper leaves, the bases of which extend around the stem, forming a cup that collects rainwater. Native to Europe, Wild Teasel is a weed of waste places throughout our area; Jun.–Oct.

Dried heads of this plant have long been used in textile mills to tease (i.e., raise the nap on cloth); thus "teasel."

Large Houstonia
Hedyotis purpurea

Bluet
Hedyotis caerulea

Princess Tree
Paulownia tomentosa

Wild Teasel
Dipsacus sylvestris

New England Aster   *Aster novae-angliae* L.
Aster Family   Asteraceae

Named after the Greek word for star, asters have distinctive flattened
flower heads. Disc flowers are yellow, changing to purple; rays vary from
white to blue or purple (rarely pink). Most bloom in late summer or fall.
Gray's Manual lists 68 species, most of which are found in our area. Because
of the number and frequent hybridization among them identification to species is often difficult.

This tall (3–7 ft.) aster has numerous violet (or rose) rays and sticky
bracts beneath. The untoothed, sessile, and lanceolate leaves are attached alternately to the hairy stem. It grows in moist, often forested, upland areas
throughout the Appalachians; Aug.–Oct.

Indians used its roots to treat fevers and diarrhea.

These asters also have untoothed (or nearly so) lanceolate leaves: Late
Purple Aster (*A. patens*), whose leaf bases almost encircle the stem, Aug.–
Oct.; Smooth Aster (*A. laevis*), whose leaf bases clasp the stem, Aug.–Oct.

Purple-stemmed Aster   *Aster puniceus* L.
Aster Family   Asteraceae

This is one of several asters with light blue or pale violet rays and sharply
pointed, lanceolate leaves. Highly variable, the leaves may be toothed or not,
and the stems are purple or green with purple tinges (as seen here). It is
mostly a plant of moist lowlands; its range extends throughout our area;
Aug.–Nov.

Stiff Aster   *Aster linariifolius* L.
Aster Family   Asteraceae

One of several asters with linear leaves, Stiff Aster is a small plant (1 ft.
tall) often seen on rocky banks or ledges. The numerous, short leaves are
stiff. Each of the one to few flower heads is terminal. It occurs from New
Bruns. and New Eng. southward; Sep., Oct.

These asters also have linear leaves: Bog Aster (*A. nemoralis*), of northern
bogs, has leaves gradually reduced in length toward the top of the stem,
Aug., Sep.; Bushy Aster (*A. dumosus*) is a tall, bushy aster bearing dozens of
flower heads, Aug.–Oct.

New England Aster
Aster novae-angliae

Purple-stemmed Aster
Aster puniceus

Stiff Aster
Aster linariifolius

Bull Thistle
Aster Family

Cirsium vulgare (Savi) Tenore
Asteraceae

Thistles are generally coarse, aggressive plants with prickly stems, prickly leaves, or both. Most species of our area are assigned either to the genus *Carduus* or *Cirsium*.

Bull Thistle is a tall (to 6 ft.) alien weed with prickles on the stems and leaves. Flower heads are large (1–2 in. across). The presence of yellow-tipped spines on the bracts (under flower heads) distinguishes it from other thistles. It is found throughout the Apps; Jun.–Sep.

These thistles also have prickles on their stems as well as leaves: Pasture Thistle (*C. pumilum*) has larger heads and lacks wings on stem; Welted Thistle (*C. crispus*) has numerous smaller heads; Jun.–Sep. (both).

Nodding Thistle
Aster Family

Carduus nutans L.
Asteraceae

Also called Musk Thistle, this tall (to 5 ft.) biennial is recognized by the several rows of large, reflexed bracts that surround the base of each huge, nodding head (3–4 in. across). Both stems and leaves are prickly. A European plant that has become an agricultural pest, it occurs in roadsides and fields especially at lower elevations of the s. Apps; Jun.–Oct.

The pappus (fluffy down attached to seeds) is used by the Yellow Warbler to line its nest.

Chicory
Aster Family

Cichorium intybus L.
Asteraceae

The sky-blue rays of the flat flower heads are attached directly to the 3–4 ft. stalks. Deeply cut leaves, which resemble those of dandelion, are mostly basal. This is an alien plant that grows in Appalachian waste places, especially roadsides at lower elevations; Jul.–Oct.

The roasted ground roots of this cultivated plant impart the characteristic taste to New Orleans–style coffee.

Bull Thistle
Cirsium vulgare

Nodding Thistle
Carduus nutans

Chicory
Cichorium intybus

Mistflower   *Eupatorium coelestinum* L.
Aster Family   Asteraceae

This, our only blue *Eupatorium*, is a 1- to 3-ft.-tall plant with flat-topped clusters of fuzzy flowers and opposite leaves. Often cultivated, it occurs naturally in moist places, often along stream banks, of the middle and s. Apps as far south as TN and SC; Aug.–Oct.

See other *Eupatorium* species in the white and the pink sections.

Rough Blazing-star   *Liatris aspera* Michx.
Aster Family   Asteraceae

Blazing-stars are tall wandlike plants with purple or pink flower heads. Leaves are alternate, linear, and often quite narrow. Identification to species usually requires close examination of the scaly bracts under the flower heads.

Rough Blazing-star grows 1–2½ ft. tall. Its flower heads are attached to the stem directly or on very short stalks. Note the floral bracts, which are rounded and slightly curled backward. It occupies sandy and rocky soils from OH and PA s. to TN, NC, and SC; Aug., Sep.

These species have stalked flower heads: Scaly Blazing-star (*L. squarrosa*), with sharply pointed scales that bend outward, Jul., Aug.; and New England Blazing-star (*L. scariosa*), which has wide, blunt-tipped bracts, Aug., Sep.

Dense Blazing-star   *Liatris spicata* (L.) Willd.
Aster Family   Asteraceae

Also called Sessile Blazing-star, this plant closely resembles *L. aspera* (*above*). It can be distinguished from it by its floral bracts, which are purple (or purple-tinged) and bluntly pointed. Growing to 5 ft. in height, it occupies wet meadows from PA s. to GA; Jul.–Sep.

266

Mistflower
Eupatorium coelestinum

Rough Blazing-star
Liatris aspera

Dense Blazing-star
Liatris spicata

Pale Purple Coneflower   *Echinacea pallida* Nutt.
Aster Family   Asteraceae

Echinacea species are prairie plants that occur only sporadically east of the Mississippi R.

Pale Purple Coneflower bears flower heads with narrow, lavender, drooping rays. Leaves, mostly basal, are long and lanceolate. Never common, it occurs at lower elevations of middle and s. Apps; Jun., Jul.

Purple Coneflower (*E. purpurea*) has darker, wider, less-reflexed rays and wider, lanceolate leaves attached to the stem; Jun.–Oct.

New York Ironweed   *Vernonia noveboracensis* (L.) Michx.
Aster Family   Asteraceae

Ironweeds are tall (to 9 ft.) plants with narrow, alternate leaves and purple flowers in clusters. "Ironweed" refers to their tough stems. New York Ironweed has narrowly lanceolate, serrate leaves with sharp tips. Flowers are violet; bracts beneath are hair-tipped. It is common in wet fields from MA and NY southward; Aug.–Oct.

Tall Ironweed (*V. gigantea*) is distinguished by its blue purple flowers, blunt-tipped bracts, and leaves that are downy beneath; Aug.–Oct. Leaves of Appalachian Ironweed (*V. glauca*) are bluish green above and paler beneath; also they are sharply pointed on both ends; Jun.–Sep.

Great Burdock   *Arctium lappa* L.
Aster Family   Asteraceae

Burdocks are large (2–8 ft. tall), coarse weeds with lanceolate leaves and bristly flower heads. From them are formed burs (fruits) that disperse their seeds by clinging to the fur of animals or people's clothing. Introduced from Eurasia, they occupy roadsides and other disturbed waste places.

Great Burdock is a large plant (to 8 ft. tall) characterized by large (1–1½ in. wide) flower heads on long stalks. It occupies low elevations of the Appalachians from Que. to PA; Jul.–Oct.

Other burdocks with smaller flower heads on short stalks include Common Burdock (*A. minus*) and Woolly Burdock (*A. tomentosum*); Jul.–Oct. (both). The latter can be recognized by the fine, matted hairs on its bracts.

Pale Purple Coneflower
Echinacea pallida

New York Ironweed
Vernonia noveboracensis

Great Burdock
Arctium lappa

Green/Brown

Jack-in-the-pulpit (Ariséma Rouge-Foncé) *Arisaema triphyllum*
(L.) Schott.
Arum Family   Araceae

Arums are characterized by a thick flowering stalk, the spadix, surrounded by a cylindrical spathe.

This striking wildflower, 12–18 in. tall, has a spadix, or "jack," and a spathe, usually striped with a flap above, that forms the "pulpit." Each plant has 1 or 2 compound leaves, each with 3 leaflets. It is found in shaded, rich woods from s. Que. southward; Apr.–Jul.; fruits, Aug.–Oct.

Green Dragon (*A. dracontium*) is a taller, more elongated plant with a spadix that extends several inches beyond the slender spathe; May–Jul.

Skunk-cabbage (Chou Puant) *Symplocarpus foetidus* (L.) Nutt.
Arum Family   Araceae

The late winter–early spring appearance (as seen here) of this 1- to 3-ft.-tall plant is quite distinctive: inside each rounded, green or purple spathe is a fleshy spadix bearing both male and female flowers. During the summer Skunk-cabbage is recognized by its huge, broadly lanceolate leaves with their bases near the ground. Emerging buds, 4–6 in. tall, can be seen in fall. It is found in swamps and along slow-moving streams from Que. s. to e. TN (at higher elevations in the south); Feb.–May.

Like many other arums, Skunk-cabbage generates heat (its temperature can be 30 degrees F above that of the air) by increasing its metabolic rate. Thus it can emerge through snow; the heat also aids in the release of aromas that attract pollinating flies and beetles.

Sweetflag *Acorus calamus* L.
Arum Family   Araceae

The spadix seen here, 2–4 in. long, bears numerous yellowish green, diamond-shaped flowers. The spathe that surrounds the spadix of most arums is lacking in this species. Narrow, swordlike leaves are 1–4 ft. long. It is a Eurasian wetland plant, often found in large colonies from Que. s. to TN and NC; May–Aug.

American Bur-reed (Rubaniera) *Sparganium americanum* Nutt.
Bur-reed Family   Sparganiaceae

Bur-reeds, not to be confused with true reeds (grass family), are aquatic plants. They have pistillate flowers that become burlike fruits. In this species, the fruits are ¾–1 inch across; May–Aug.

Also found in shallow water, Bur-reed (*S. androcladium*) has fruits more than 1 in. in diameter; May–Sep.

Both are found sporadically throughout the Apps.

Jack-in-the-pulpit
Arisaema triphyllum

Skunk-cabbage
Symplocarpus foetidus

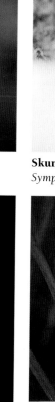

Sweetflag
Acorus calamus

American Bur-reed
Sparganium americanum

River Oats   *Chasmanthium latifolium* (Michx.) Yates *[Uniola latifolium]*
Grass Family   Poaceae

One must often look closely to see the florets (tiny flowers) of grasses. They have greatly reduced, inconspicuous sepals and petals, a feature common in species that rely on wind pollination. The fruits of cultivated grasses, such as corn, wheat, and rice, are grains that help to feed the world. In addition to their importance as food plants, grasses also help to hold soil. They occupy a wide spectrum of soil conditions from moist and cool to dry and hot.

This graceful plant, 2–4 ft. tall, bends with the weight of 8- to 20-in.-long flat spikelets. River Oats, or Wild Oats, is common along shaded streams, pond margins, and other wet places in the mountains from PA and WV southward; Jun.–Oct.

Two other *Chasmanthium* species of our area have fewer than 8 spikelets: *C. laxum* has smooth sheaths covering the spikelets, Jun.–Oct.; *C. sessiliflorum* has hairy sheaths, Aug.–Oct.

Eulalia   *Miscanthus sinensis* Andersson
Grass Family   Poaceae

This species, showier than most grasses, is a clumped perennial that reaches 6–8 ft. in height. Brought from e. Asia as an ornamental, it often escapes and is seen in old fields and along roads throughout much of middle and s. Appalachia; Sep.–Nov.

Wild Oatgrass   *Danthonia compressa* Austin
Grass Family   Poaceae

Wild oatgrasses (*Danthonia* species) are tufted perennials with narrow leaves and small terminal racemes or panicles. This 1-ft.-tall species, also called Allegheny Fly-back, is found in open woods and meadows from s. Que s. to the mountain balds of TN and NC; Jun.–Aug.

River Oats
Chasmanthium latifolium

Eulalia
Miscanthus sinensis

Wild Oatgrass
Danthonia compressa

Nut-grass (Amande de Terre)   *Cyperus esculenta* L.
Sedge Family   Cyperaceae

"Sedges have edges" is a means of telling sedges from grasses (*above*); whereas the latter have round, hollow stems, those of sedges are typically solid and triangular (as seen in cross-section). Most of the 200–300 Appalachian sedge species are found in wet places. Other sedges are featured in the white section.

Nut-grass belongs to a group known as umbrella sedges; a cluster of compact, brown flower heads are surrounded by a whorl of leaves. It is 6–30 in. tall and has minute flowers borne in scaly heads. Native to tropical Amer., it is found in damp, sandy soils of the Apps n. to s. Que.; Aug.–Oct.

Narrowleaf Cat-tail (Masette)   *Typha angustifolia* L.
Cat-tail Family   Typhaceae

Cat-tails are tall plants with compact cylinders of tiny flowers that grow in shallow water. The staminate (male) flowers are located above the pistillate (female) ones. Leaves of Narrowleaf Cat-tail are grasslike, and there is a gap between the two types of flowers. Originally from Eurasia, it grows in alkaline waters at low elevations of the Apps from s. Que. southward but is more common in adjacent lowlands; May–Jul.

Common Cat-tail (*T. latifolia*) is seen more often but occurs in similar habitats. It has wider leaves and lacks the flower gaps; May–Jul.

Several parts of cat-tails are edible: the young spiny shoots can be used like asparagus; the pollen, as flour; and the rootstocks, as a starch source in winter.

False Hellebore (Tabac du Diable)   *Veratrum viride* Ait.
Lily Family   Liliaceae

This tall (4–8 ft.), stout Appalachian plant is quite distinctive. Also called Indian Poke, it grows in open moist woods and bogs at increasingly higher elevations from Que. to GA; Apr.–Jul.

The s. Appalachian plant *V. parviflorum*, also called False Hellebore, has leaves that are primarily basal and smaller green flowers that form less conspicuous inflorescences; Jul.–Sep.

False hellebores have been used to treat various heart ailments. They contain highly poisonous alkaloids (as indicated by their French name, which means "devil's tobacco").

Nut-grass
Cyperus esculenta

Narrowleaf Cat-tail
Typha angustifolia

False Hellebore
Veratrum viride

False Aloe　　　　　　　　*Agave virginica* L. *[Manfreda virginica]*
Agave Family　　　　　　　　　　　　　　　　　　　Agavaceae

Thick leaves form rosettes at the base of these tall (3–6 ft.) flowering stalks. Note the inconspicuous flowers; they are believed to be pollinated by moths attracted to the scent released at night. It is found in dry sandy soils of the s. Apps; Jun., Jul.

Preparations from the roots were used by Native Americans in the treatment of dropsy, diarrhea, and other ailments.

Crane-fly Orchid　　　　　　　*Tipularia discolor* (Pursh) Nutt.
Orchid Family　　　　　　　　　　　　　　　　　　Orchidaceae

The 1- to 2-ft.-tall spikes of this inconspicuous plant appear long after the spiny leaves have withered. The tiny flowers are purplish brown. It occurs at low to middle elevations of the Apps in moist forests from MA and NY southward; Jun.–Aug.

Putty-root　　　　　*Aplectrum hyemale* (Muhl. ex Willd.) Nutt.
Orchid Family　　　　　　　　　　　　　　　　　　Orchidaceae

The single, leathery basal leaf at the base of each flowering stalk is the most useful means of identifying this wildflower. As the leaf withers, the flowering stalk (1 ft. tall or less) develops. Small flowers combine shades of brown, green, and white, sometimes violet. It is a common perennial of moist, rich woods from s. Que. to GA; May, Jun.

"Putty root" refers to a sticky paste, sometimes used to mend pottery, which can be produced from the roots.

Northern Green Orchis　　　　　*Habenaria hyperborea* (L.) R. Br.
　　　　　　　　　　　　　　　　　　[Platanthera hyperborea]
Orchid Family　　　　　　　　　　　　　　　　　　Orchidaceae

This tall (1–2½ ft.) orchid consists of a stout stalk with tightly packed green or yellow green flowers. Just below the flower spike are several sharply pointed, lanceolate leaves. It is principally a bog plant of Que. and New Eng.; May–Jul.

Long-bracted Orchis (*H. viridis*) is similar but has wider, more numerous leaves that continue to the top of the stalk; Jun.–Aug. Small Woodland Orchis (*H. clavellata*) is a smaller plant with a nearly naked flower stalk (only a single large leaf is attached at its base); Jul., Aug.

False Aloe
Agave virginica

Crane-fly Orchid
Tipularia discolor

Putty-root
Aplectrum hyemale

Northern Green Orchis
Habenaria hyperborea

Little Brown Jug *Hexastylis arifolia* (Michx.) Small *[Asarum arifolia]*
Birthwort Family Aristolochiaceae

This plant can be distinguished from Wild Ginger (red section) by its brown (vs. maroon) juglike flowers, which lack extended calyx lobes, and its evergreen leaves (3–4 in. long), which are thick and leathery. Extending no farther north than WV, it grows in upland ravine slopes and woods; Apr., May.

Catawba Indians used the root of this plant to treat heart pains.

Dutchman's-pipe *Aristolochia macrophylla* Lam. *[A. durior]*
Birthwort Family Aristolochiaceae

This woody vine often climbs 100 ft. or higher into tree tops. The cordate leaves are very large. The flowers are interesting, though not conspicuous, consisting of 3 short calyx lobes that form the end of the "pipe." It occurs naturally in moist woods from s. PA southward and is commonly cultivated north of its natural range as an ornamental to shade porches; Apr.–Jul.

Virginia Snakeroot (*A. serpentaria*) is a smaller (8–10 in.), erect plant with narrower leaves and similar but smaller flowers. It ranges farther north than *A. macrophylla*; May, Jun.

The roots of Virginia Snakeroot were widely used by Indians to treat snakebite and a variety of other ailments. It was listed in various pharmacopeia until the mid-20th century. Plants of both species contain aristolochic acid, an antitumor compound.

False Nettle *Boehmeria cylindrica* (L.) Swartz
Nettle Family Urticaceae

Included in this family are coarse weeds of waste places. The simple leaves have toothed margins. The small, greenish flowers occur in spreading or drooping clusters.

False Nettle is a 1- to 3-ft.-tall perennial; note its lanceolate leaves with dentate margins. It occupies moist lowland habitats from s. Que southward; Jul.–Oct.

Stinging Nettle (*Urtica dioica*) is a similar plant native to Europe but found throughout our area. It has a hollow, square stem covered with large, bristly hairs. The hairs, on contact with skin, cause an intense burning sensation for up to an hour. Nevertheless, these plants have been used in folk medicine, and modern research indicates their usefulness in the treatment of asthma and prostate cancer; Jun.–Sep.

Little Brown Jug
Hexastylis arifolia

Dutchman's-pipe
Aristolochia macrophylla

False Nettle
Boehmeria cylindrica

Green Violet   *Hybanthus concolor* (Foster) Sprengel
Violet Family   Violaceae

Although a member of the violet family, this plant bears little resemblance to violets. The small (¼ in. long) flowers are arranged singly in the axils of the lanceolate leaves. Green Violet grows to a height of 2–3 ft. in moist, basic soils from the Adirondacks and New Eng. southward; Apr.–Jun.

Blue Cohosh (Caulophylle Faux-Pigamon)   *Caulophyllum thalictroides*
                                                              (L.) Michx.
Barberry Family   Berberidaceae

This distinctive smooth perennial has bluish green leaves divided into 3 leaflets. Small green or yellow flowers are arranged in a terminal raceme; berries are dark blue. It is common in rich woods from New Bruns. to SC and TN; Apr.–Jun; fruits, Jul.–Sep.

Native Americans pulverized the roots and used them to treat rheumatism, bronchitis, and menstrual cramps; they were also used as an aid in childbirth. Modern herbalists use the roots to treat these same ailments. There is scientific evidence that the plant is effective against rheumatism and that it has antispasmodic properties, but the berries are poisonous.

Alpine Willow   *Salix uva-ursi* Pursh
Willow Family   Salicaceae

This is one of several species of dwarfed, prostrate willows characteristic of alpine areas of the n. Apps. Seen here thriving on thin soil over granite are female plants with their small (1 in. long) flowering heads; male heads are lighter in color, soft, and fuzzy. North of our area it is widespread in the arctic tundra. In the Apps it is confined to alpine areas such as those of the Adirondacks and White Mountains and, less often, those of ME and Que; Jun., Jul.

The bark of the European White Willow (*S. alba*) has been used for centuries as a source of salicin, a painkiller related to aspirin. Indians also used Black Willow (*S. nigra*), a native tree, as a painkiller and astringent.

Green Violet
Hybanthus concolor

Blue Cohosh
Caulophyllum thalictroides

Alpine Willow
Salix uva-ursi

Mountain Maple (Erable à Epis) *Acer spicatum* Lam.
Maple Family Aceraceae

Maples are our only trees with opposite, simple (but lobed) leaves. Their key-shaped, winged fruits (samaras) in pairs are also distinctive.

Mountain Maple is a small (to 30 ft. tall) understory tree. Its leaves, 5–6 in. long, have 3 lobes and coarsely toothed margins. Note the yellowish green flowers in erect racemes. Samaras change during a season from yellow to red to brown. More common in the n. Appalachians, it is found at increasing elevations to s. GA; May–Jul.

Striped Maple (*A. pensylvanicum*), a similar tree, can be distinguished by its green bark with white vertical stripes and its flowers arranged into slender, drooping racemes; Jun.–Sep.

Wafer-ash *Ptelea trifolia* L.
Rue Family Rutaceae

A misnamed plant (not related to ashes, *Fraxinus*), this aromatic shrub or small tree (10–20 ft. tall) has leaves with 3 leaflets. Small green flowers are in cymes; the waferlike samaras (fruits) that follow are flat, circular, and 2-seeded. Wafer-ash is found along shaded stream banks and rocky slopes. It is scattered throughout the Apps but is less common northward. The leaves of Wafer-ash resemble those of Poison Ivy but can be distinguished by noting that the terminal (middle) leaflet tapers at its base, rather than being attached by a stalk as in Poison Ivy; Apr.–Jul.

Cancer-root *Conopholis americana* (L.) Wallroth
Broom Rape Family Orobanchaceae

Giving the initial impression of a pine cone, this brown parasitic plant has thick, unbranched, 4- to 6-in.-long stems. The small, hardly noticeable, white, bilabiate (2-lipped) flowers are arranged in axils of the scales (reduced leaves). Lacking chlorophyll, its roots form specialized attachments (haustoria) to roots of deciduous trees by which it receives nutrition. Cancer-root occurs in deciduous forests throughout the Apps; Apr.–Jun.

There are two other brownish parasitic plants of this family in the Appalachians. Beech-drops (*Epifagus virginiana*) is a taller, more slender branching plant with white flowers; it parasitizes roots of beech trees; Aug.–Oct. One-flowered Cancer-root (*Orobanche uniflora*) is a delicate plant that produces above ground several slender flower stalks (5–6 in. tall), each of which bears a single, white tubular flower with 5 terminal lobes. It parasitizes various deciduous trees; Apr., May.

Mountain Maple
Acer spicatum

Wafer-ash
Ptelea trifolia

Cancer-root
Conopholis americana

American Columbo   *Frasera caroliniensis* Walt. *[Swertia caroliniensis]*
Gentian Family   Gentianaceae

A long-lived perennial, American Columbo produces a basal rosette of large (to 1 ft. long) oblong leaves for several years before "bolting" (forming a flower stalk), after which the entire plant dies. The stalk may reach a height of 6 ft. or more and bear dozens of flowers such as seen here. Note the purple spots on the greenish petals; flowers are about 1 in. across. This striking plant is not common but is found in rich woods and rocky limestone soils from NY southward; Jun., Jul.

Summer Grape   *Vitis aestivalis* Michx.
Grape Family   Vitaceae

What mountain youngster has not used a high-climbing grapevine to swing on a hillside or into a chilly stream? The Appalachian flora includes nearly a dozen *Vitis* species; most are quite similar to the one featured here. Leaves are more or less lobed, and the bark is shreddy. Early summer flowers, greenish and in panicles, are followed by autumn fruits, which are eaten by both humans and wildlife.

Leaves of Summer Grape are 3–6 in. long and more deeply lobed than those of other species. Dark purple fruits are a little less than ½ in. across. This species is quite common in open forests and along streams from New Eng. southward; May, Jun.; fruits, Sep., Oct.

Muscadine (*V. rotundifolia*) is distinct from other grapes: leaves are rounded with large teeth; bark doesn't readily shed; and fruits are larger and have a thick skin; May, Jun.; fruits, Sep., Oct.

Rugel's Ragwort   *Cacalia rugelia* (Shuttleworth ex Chapman)
Barkley & Cronquist *[Senecio rugelia]*
Aster Family   Asteraceae

This 1- to 2-ft.-tall plant lacks rays around the less-than-showy tan flower heads. The large (8–10 in. long) basal leaves are widely cordate. The only plant known to be endemic to the Great Smoky Mountains National Park, it is often locally abundant in openings of spruce-fir forests; Jun.–Sep.

The more common and widespread Indian Plantain (*C. atriplicifolia*) has smaller, rayless, white-flowered heads in flat-topped clusters. Its fan-shaped leaves have coarsely toothed margins; Jun.–Sep.

American Columbo
Frasera caroliniensis

Summer Grape
Vitis aestivalis

Rugel's Ragwort
Cacalia rugelia

Glossary

Abscission layer. Separation layer that forms at the base of the petiole (stalk) of a leaf; its formation causes the leaf to be shed.

Alfisols. Soils associated with deciduous forests; typically they are brown or gray brown.

Alluvial. A type of soil resulting from the deposition of sediments transported by water.

Alpine plants. Plants found above timberline on high mountains.

Alpine tundra. The ecosystem above timberline of high mountains.

Angiosperms. The largest group of plants (250,000 described species); they reproduce by flowers, fruits, and seeds.

Annual. A plant that lives only a single growing season.

Arctic tundra. The cold, treeless, prairie-like ecosystem found north of the boreal forest.

Azonal soils. Soils lacking horizontal layers; those that were carried to their present site by wind, water, or gravity.

Bald (mountain). Open treeless areas, especially at high elevations of the southern Appalachians where forests would be expected.

Beech gaps. In the southern Appalachians, areas composed of stunted American beech trees where these trees interface with the spruce-fir forest above.

Biodiversity. The total number and variety of species of a particular ecosystem or biotic community.

Biome. A large geographical area dominated by one principal type of ecosystem; e.g., prairie, deciduous forest, or tundra.

Biosphere. The entire Earth where life exists along with the soil, air, and water that support life.

Biota. The collective life of an area or ecosystem; includes plants, animals, and microorganisms.

Biotic community. All the living organisms of a local area or ecosystem.

Bog. A wetland ecosystem in which acidic conditions, resulting from a large amount of peat, exist.

Boreal forest. A forest composed primarily of evergreen conifers such as spruce, pine, or fir; winters are generally long and cold.

Calcareous. Refers to basic soils formed from calcium-containing rocks, especially limestone.

Canopy. The uppermost layer of tall trees in a forest.

Circumboreal. Refers to the range of species found in the northern parts of North America, Europe, and Asia.

Climax ecosystem. In traditional ecological theory, a stable ecosystem in equilibrium with natural factors, especially climate.

Conifer. A tree, usually evergreen, that produces seeds within cones.

Continental drift. The geological theory that recognizes that continents move slowly across the face of the earth.

Cove. Term used, especially in the southern Appalachians, for a valley protected on several sides by surrounding mountains.

Deciduous forest. A forest in which most of the trees lose their leaves each fall.

Devonian. Geological time period extending from about 405 to 360 million years ago.

Dicot. One of the two major groups of flowering plants; characterized by flower parts in 4s or 5s and net-veined leaves. *See also* monocot.

Dioecious. Refers to plants with staminate (male) and pistillate (female) flowers on separate individuals. *See also* monoecious.

Dominant species. One or more species of an ecosystem or biotic community that, because of their size or numbers, exert a major influence.

Drupe. A fruit with a pit that contains a seed inside (e.g., cherry).

Ecology. The branch of biology concerned with interactions among organisms and their environments.

Ecosystem. A local unit of nature including both living and nonliving (soil, water, light, temperature) factors; e.g., pond, coral reef, bog, field, forest.

Ecotone. A transitional zone between two adjacent biomes or ecosystems.

Endemic. A plant or animal species whose distribution is limited to a particular local area.

Entisols. Immature soils lacking stratification (recognized layers) and often associated with a disturbance.

Exotic. A plant or animal occurring outside its native area.

Fauna. The collective animal life of an area.

Flora. The collective plant species of an area.

Floristic. Referring to the total listing of plants of an area without regard to their abundance.

Forest stand. A particular forest with boundaries (as opposed to forests in the abstract).

Generic name. The first part of a scientific name; e.g., *Iris* is the generic name for *Iris cristata*.

Heath. A member of the Ericaceae, a family of flowering plants comprised mostly of shrubs.

Heliophyte. Plants that thrive in habitats with high light intensity.

Herb. A plant with nonwoody tissues (as compared to shrubs and trees with woody tissues); also applied to any plant used for a particular practical use, such as for medicine, dyeing, or flavoring.

Horizons (soil). Horizontal strata of soils as seen in a soil profile (vertical cut).

Hydric. Refers to a wet habitat or environment.

Inflorescence. A cluster of individual flowers on a single stalk (peduncle).

Mesic. An environmental condition intermediate between wet (hydric) and dry (xeric); i.e., moist.

Mesophyte. A plant adapted to mesic conditions.

Microclimate. The climate of a particular local site such as a cave or the south side of a hill; each ecosystem possesses numerous microclimates.

Microhabitat. A distinctly different portion of a more general habitat.

Monocot. One of the two major groups of flowering plants; characterized by flower parts in 3s and parallel-veined leaves. *See also* dicot.

Monoecious. Refers to plants with both staminate (male) and pistillate (female) flowers on each individual. *See also* dioecious.

Naturalist. One with a broad interest in or knowledge of organisms and natural habitats.

Old field. An abandoned crop field in the process of ecological succession; often recognized by its characteristic weeds and shrubs.

Old-growth forest. A forest, although not necessarily a virgin one, that approximates a climax forest.

Orchards. In the southern Appalachians, a naturally occurring forest consisting of stunted trees somewhat suggestive of a cultivated orchard.

Organism. Any individual living thing, be it plant, animal, or microorganism.

Paleozoic Era. Geological era of 590 to 248 million years ago during which there was a great diversification of life, especially multicellular animals.

Peat. The highly organic material resulting from the decomposition of plants, especially sphagnum moss.

Perennial. Plants that last three or more growing seasons, typically reproducing many times.

pH. A scale used to indicate the acid/alkaline condition of a solution or the soil; numbers below 7 indicate varying degrees of acidity; those above 7 indicate alkalinity.

Pioneer plant. One that invades recently disturbed areas, thus initiating ecological succession.

Plate tectonics. The geological theory that postulates the existence of a dozen or so movable plates at the surface of the earth.

Pleistocene. Ice Age, beginning about 2 million years ago, during which glaciers covered the northern part of North America.

Prairie. A naturally occurring ecosystem dominated by grasses and related plants.

Reflexed. Refers to plant parts, especially sepals or petals, that are bent downward toward the stem.

Refugium. A local area that, because of its favorable microclimate, serves as a refuge for plants and animals during a period of unfavorable conditions existing over a wider area.

Rhizome. Underground, usually horizontal stem of a plant.

Saponin. Chemicals, produced by some plants, that form a lather.

Secondary forest. A forest that has regrown following cutting or other disturbances.

Saprophytic. Refers to a plant that lives on dead or decomposing organic material.

Sessile. Refers to a leaf with its blade attached directly to a stem without a petiole.

Species. A particular kind of plant, animal, or microorganism; members of a given species typically breed within the group but not outside it.

Spodosols. Soils typically associated with boreal forests; they tend to be thick and highly acidic.

Successional ecosystems. Ecosystems in the process of undergoing ecological succession.

Taxonomy. The branch of biology concerned with naming and classifying species of living things.

Tepals. Sepals and petals collectively.

Timberline. The uppermost extent of trees on high mountains (above is an alpine tundra).

Tundra. Areas of low-growing vegetation characteristic of very cold areas; included are both the arctic tundra of the north and the alpine tundra at the tops of high mountains.

Ultisols. Soils typical of southeastern U.S.; they are typically reddish and have a high clay content.

Vegetation. All the collective plant life of a local area.

Virgin forest. A climax forest that has never been disturbed.

Weed. An organism, usually a plant, that thrives in disturbed sites; a plant growing in a place where it is not desired.

Xeric. Refers to a dry habitat or environment.

Mountain Natural Areas

Described here are some of the diverse natural areas where wildflowers and associated biota can be experienced. The number of excellent areas to visit is far too large to include them all; the ones described here, arranged by mountain regions, are representative.

Appalachian Trail and Scenic Drives

As first proposed in the 1920s, the Appalachian Trail (AT), the world's first "linear park," was to have extended from Mt. Washington in New Hampshire to Mt. Mitchell in North Carolina. Such a route exists today with a northern extension to Maine's Mt. Katahdin and a southern one terminating atop Georgia's Springer Mountain.

Having hiked portions of the AT, I can understand at least partially, the special feelings of "through-hikers," those who devote the 6 months necessary to walk the entire length of the trail. I can also appreciate the care with which numerous local trail clubs along its length take responsibility for maintaining their section of the trail. Along the trail, you can observe not only the pattern of vegetation that changes with latitude, elevation, and aspect (north- vs. south-facing slopes) but also the geology and important historical sites.

Extending 105 miles south from Fort Royal, Virginia, and often crisscrossing the AT is Skyline Drive. Designed to take advantage of spectacular views from the crests of the Blue Ridge Mountains, it bisects Shenandoah National Park. Near Waynesboro, Virginia, Skyline Drive connects with the Blue Ridge Parkway, which continues its twisting course along the spine of the mountains to its southern terminus, the Great Smoky Mountains National Park (GSMNP) in North Carolina. Because of its higher rainfall and greater elevational range, the North Carolina segment of the parkway is richer botanically than farther north. At the highest elevations can be found good examples of heath balds. Total length of the Blue Ridge Parkway is 469 miles. Don't plan to make good time along these drives: the speed limit is generally 45 miles per hour, and there are notable historical attractions as well as scenic overlooks and opportunities for botanizing around every curve. Also there are strategically located visitor centers where information, maps, and books are available.

Northern Appalachians

QUEBEC

Much of southeastern Quebec, particularly the Gaspé Peninsula, lies within Appalachia. Described below are several natural areas along Hwy. 132, the perimeter road for picturesque Gaspésia.

Les Jardin de Metis. This provincial garden, about 6 miles (10 km) east of Saint-Flavie, possesses a favorable microclimate, making it possible for many plants to grow here that are otherwise found much farther south. Although many of the plants are cultivated ornamentals, there are alpine plants and many other native Appalachian species. Open daily June through mid-September. Admission charged.

Parc de la Gaspésia. Here, near the town of Sainte-Anne-des-Monts, are located the Chic-Chocs, the highest peaks of southern Quebec. From the interpretation center, you may hike a trail that leads you to the summit of Mt. Albert; you will pass through a boreal forest to alpine tundra. A trail map available at the center will indicate the location of serpentine soil areas with their botanical rarities. Within this immense park are dining, lodging, and camping facilities. Plan to spend several days if possible.

Forillon National Park. This park, near the city of Gaspé at the tip of the peninsula, shelters 3 types of habitats: sand dunes, salt marshes, and limestone cliffs with arctic-alpine plants. Obtain at a visitor center the booklet that describes (in French or English) the 17 species of rare plants known to inhabit these cliffs. Admission charged.

Parc de l'Ile-Bonaventure. Near the resort town of Perce is this island with its diverse plant life, including conifers and arctic-alpine plants. No admission to the park but a charge for the ferry from Perce; open June–August.

NEW ENGLAND AND NEW YORK

Especially in the northern and western portions of this 7-state region are numerous natural areas with their varied forests, wetlands, and flora. Several mountains have treelines.

Baxter State Park/Mt. Katahdin (Maine). This mountain, the highest point (5,268 ft.) in Maine, is within Baxter State Park, a 200,000-acre wilderness in the northcentral part of the state. At the top is an alpine environment with flora similar to that on Mt. Washington. You may reach the summit by a 5-mile climb from the nearest road. Many areas of botanical interest are also at lower elevations. Near the northern entrance, Lower Fowler Pond is recommended. From the south gate, take the road (5 mph) to Daicey Pond; the nature trail that encircles it takes you through diverse terrestrial and aquatic habitats (maps and checklists available). Cabins and lean-to campsites are available. There is a substantial entrance fee for out-of-staters, despite the rough roads and primitive facilities.

Acadia National Park (Maine). More accessible to more people than the remote north woods, this park is located largely on Mt. Desert Island southeast of Bangor and near the resort town of Bar Harbor. Numerous opportunities abound for botanizing, especially on the loop road, but 2 sites are especially recommended: Mt. Cadillac and the Wild Gardens of Acadia at Sieur de Mont Spring. A paved spur road leads to the top of Mt. Cadillac, which, at 1,530 ft., is the highest point along the U.S. Atlantic Coast. Besides a view of the ocean, lakes, and other nearby mountains, there are numerous wildflowers such as Rhodora, Three-toothed Cinquefoil, and Mountain-cranberry among the granite boulders. The Wild Gardens of Acadia contains 400 species, virtually all those of the park. Separate sections include plants of deciduous woods; roadsides and meadows; bogs, marshes, and ponds; coniferous woods; mountains; and dry rocky places. A checklist is available at the adjacent nature center.

White Mountains/Mt. Washington (New Hampshire). Together with Mt. Adams, Mt. Jefferson, and Mt. Monroe, Mt. Washington is a part of the Presidential Range of northern New Hampshire's White Mountains. At 6,288 ft., it is the highest mountain in northeastern North America. You may reach the summit by the Mt. Washington Auto Road, the Mt. Washington Cog Railway, or the AT. These and other peaks are linked by the Appalachian Mountain Club hut system, which includes 8 lodges a day's hike apart.

Berkshires/Mt. Greylock (Massachussets). The Berkshire Mountains occupy the westernmost part of Massachusetts, an area well known for its summer festivals and cultural events. But there are also areas for the naturalist; one I would recommend is Greylock State Preservation. From Pittsfield, go north on Hwy. 7; about a mile past Lanesboro, look for the turnoff sign. At the visitor center are displays and maps that will lead you the 8½ miles by a logging road to the summit of Mt. Greylock. Along the way are deciduous and mixed forests with considerable diversity of wildflowers. At the summit (at 3,491 ft. the highest point in the state) is a tower that on a good day allows you to see mountains of 5 states.

Beckley Bog (Connecticut). This wetland, located in a narrow valley of the Northwestern Highlands near Norfolk, is owned by the Nature Conservancy. An excellent example of a northern bog, it includes a 7-acre lake within a total of 600 acres. Typical bog species, including carnivorous plants and heaths, are surrounded by Black Spruce and Tamarack.

Adirondack State Park (New York). This park, now 100 years old, is the largest (9,400 square miles) of any park in the lower 48 states. It is, however, not totally state owned; within its boundaries is a mixture of public and private lands. Near the center of the park, on Hwy. 30, is the Adirondack Museum, which consists of 18 buildings; it is a good introduction to the human and natural history of the area. Northward on Hwy. 30, you can travel through a wilderness in which northern tree species increasingly dominate the forest: Red and White Spruce, Yellow and White Birch, along with Sugar Maple. Habitats and wildflowers of the Adirondacks parallel closely those of New England at similar elevations and latitudes.

Near Tupper Lake, take Hwy. 3 east to Saranac Lake, where Spring Pond Bog is located, an area of 550 acres and one of the most significant bogs in the park. There you will find large mats of the moss *Sphagnum rubellum,* which give a red color to large portions of the bog, along with many typical bog species; the Spatulate-leaved Sundew is found along pond margins.

About 20 miles east of Saranac Lake and north of Lake Placid is Whiteface Mountain (4,867 ft.), which can be reached by an 8-mile paved road. Much less accessible, but a great destination for experienced hikers, is Mt. Marcy, at 5,344 ft. New York's highest peak; it is just south of Lake Placid. Perhaps the best starting point is the town of Keene Valley. From there, one trail, about 10 miles long, travels up the mountain along John Brook. You will need to carry food and shelter for the night. Along the trail you will pass through much the same zones as described for Mt. Washington, including an excellent alpine tundra on the summit.

Middle Appalachians

SPRUCE KNOB–SENECA ROCKS NATIONAL RECREATION AREA (WEST VIRGINIA)

In the eastcentral part of the state, this area includes Spruce Knob, a mountain that, at 4,861 ft., is the highest point in the state. Much of its summit is a heath bald (see chap. 4) including such notable shrubs as Mountain Holly and Rose Azalea; Bunchberry is at its southernmost limit here. At the top is an observation tower that permits a 360-degree view; also recommended is the ½-mile Whispering Spruce Trail. Seneca Rocks (see fig. 1), with their sheer 900-foot sandstone pinnacles, is a popular site for mountain climbers. Camping is available at this area as well as on Spruce Knob.

DOLLY SODS (WEST VIRGINIA)

This natural area just north of Seneca Rocks is also part of the vast Monongahela Forest. Its name refers to the extensive open areas comprised of grasses and sedges. Its flora is very similar to that of Spruce Knob.

CRANBERRY GLADE BOTANICAL AREA (WEST VIRGINIA)

Just 50 miles southeast of Spruce Knob is this 600-acre wetland-peatland complex. Included are 4 open areas, each 8 acres or more: Big Glade, Flag Glade, Long Glade, and Round Glade. Cranberry Glade is the largest and, no doubt best, example of a sphagnum glade. Without a special permit, you are confined to boardwalks. Among the diverse flora are Buckbean and Bog Rosemary (neither found elsewhere in West Virginia), Skunk-cabbage, Grass-of-Parnassus, and several orchids and carnivorous plants.

ROSECRAN BOG (PENNSYLVANIA)

Near the town of the same name in the central part of the state, this 152-acre natural area has Highbush-blueberry, Mountain Holly, and cranberries, along with a variety of sedges and other bog herbs.

BEAR KNOB QUADRANGLE (PENNSYLVANIA)

This largely unknown state reserve is now, despite having been logged at the turn of the century, a place where you may encounter bears and coyotes. Among the flora are Mountain Laurel, trilliums, huckleberries, and other plants typical of the Al-leghenies. Bear Knob is at the almost exact center of the state; from Harrisburg, take I-80 north, then south on Hwy. 220 for 6 miles. Camping and cabins are available at nearby Black Moshannon State Park.

SHENANDOAH NATIONAL PARK (VIRGINIA)

This park extends from Front Royal to Waynesboro, the entire length of Skyline Drive. Elevations vary from 600–4,050 ft., which results in considerable floral di-versity. As in much of middle and southern Appalachia, herbaceous species are most conspicuous in April and May, with the rhododendron spectacle following in late June. Camping and lodging are available.

Southern Appalachians

BIG SOUTH FORK NATIONAL RIVER AND RECREATION AREA (KENTUCKY/TENNESSEE)

On the Cumberland Plateau with major portions in both Tennessee and Kentucky, this recently established natural area includes the most remote and primitive por-tions of the plateau. Deep canyons (600 ft.) are formed by the Big South Fork of the Cumberland River. Included in the area are some (reported) virgin forests; there are also a number of federally listed plant and animal species. The visitor's center can be reached from either Oneida or Jamestown, Tennessee, via Hwy. 297. There are nu-merous hiking and horse trails as well as several campgrounds.

SAVAGE GULF STATE NATURAL AREA (TENNESSEE)

North of Tracy City, along the western escarpment of the Cumberland Plateau, are gorges with virgin mixed mesophytic forests. These are remnants of this once more-widespread forest type of eastern North America. Flowering shrubs and herbaceous wildflowers, too, are diverse and are similar to those of the cove hardwood forests of the Great Smoky Mountains.

ROAN MOUNTAIN (TENNESSEE/NORTH CAROLINA)

This vast high country between the 2 states is bisected by 15 miles of the AT. No bet-ter place could be chosen for viewing both grassy and heath balds, with their many mountain wildflowers. A wide variety of herbaceous species flower in May and are followed in late June by the spectacular Mountain Rosebay and Flame Azaleas; also in flower at this time are Northern Sandwort, Gray's-lily and Three-toothed Cinque-foil, among many others. The highland area can be reached via Hwy. 19-E from Eliza-bethton, Tennessee, or from the Blue Ridge Parkway near Linville, North Carolina, via 194 and 19-E.

UNIVERSITY BOTANICAL GARDENS (NORTH CAROLINA)

This seminaturalistic garden is located on a 10-acre tract that borders the Asheville campus of the University of North Carolina. It contains more than 600 species of flowering plants; most are native to the mountains of the state, including some rare plants such as Oconee Bells. A useful guide available at the garden office gives locations and flowering times.

GREAT SMOKY MOUNTAINS NATIONAL PARK (NORTH CAROLINA/TENNESSEE)

If you can explore only one Appalachian area, this should be it. Here is found, in an area where north meets south, essentially all the forest types and other Appalachian habitats.

I suggest a stop first at the Sugarlands Visitor Center near the resort town of Gatlinburg, Tennessee. There is an interpretive museum and slide program, along with a bookstore where books, tapes, and maps can be obtained. Be sure not to miss Cades Cove, 20 miles southwest via scenic Little River Road; here is found a restored village as well as an excellent cove hardwood forest. Returning to Sugarlands, Hwy. 411, as you travel southeastward to Newfound Gap, takes you upward progressively some 3,600 ft. through these forest types: oak-pine, cove hardwood, northern hardwood, and spruce-fir. From Newfound Gap, Clingman's Dome can be reached by a spur road. At 6,643 ft., the highest point in the park, it has an elevated observation deck that can be reached by a short (½ mile) but steep trail. Also reachable by a 2-mile trail is Andrews Bald, the most accessible of the grassy balds of the park.

Anytime is a good time to visit the GSMNP, but 3 seasons have special attractions for wildflower enthusiasts. Spring is when the largest number of wildflowers is in bloom, especially at lower elevations from mid-April to mid-May. (Gatlinburg Wildflower Pilgrimage is traditionally the fourth weekend of April.) In summer, rhododendrons peak from late June through July. In the fall (especially October), deciduous trees and shrubs are ablaze with color.

Within the park are several developed campgrounds. Primitive camping along the Appalachian Trail takes you to locales most visitors never see; the AT is easily accessed at Newfound Gap.

HIGHLANDS BOTANICAL GARDENS (NORTH CAROLINA)

On the eastern edge of the small resort town of Highlands, some 50 miles southeast of the GSMNP, is the Highlands Biological Station with its adjacent wildflower gardens. There are several hundred species of labeled trees, shrubs, and wildflowers in a natural setting of a hemlock-deciduous forest and a mountain bog. Many plants here are rarely seen this far south. A bulletin board lists plants that are in flower. From Franklin, take Hwy. 64 east up a river gorge to Highlands.

JOYCE KILMER–SLICKROCK WILDERNESS (NORTH CAROLINA)

Here, in this 15,000-acre preserve, occurs the most extensive and pristine southern Appalachian forest outside the GSMNP. A very high rainfall contributes to a "temperate rainforest" with many lichens, mosses, and ferns, as well as a vast array of wildflowers typical of a cove hardwood forest. Especially recommended is the figure-eight trail, an easy walking trail through a virgin forest of Joyce Kilmer. The wilderness area can be reached by taking Hwy. 1127 12 miles west of Robbinsville.

BRASSTOWN BALD (GEORGIA)

In north Georgia are many natural areas, especially the Chattahoochee National Forest, which includes Brasstown Bald Mountain. From Dahlonega, take Hwy. 19 north, then east on Hwy. 180. The drive to the top is steep but paved, and along the roadside are many wildflowers, both native and nonnative. On top of Georgia's highest summit (4,784 ft.) is an interpretive center with an observation deck surrounded by American Mountain-ash and various heaths. At nearby Vogel State Park, where the Showy Orchis can be seen in masses, cabins and camping are available.

Bibliography

Abrams, M. D. 1992. Fire and the development of oak forests. *BioScience* 42: 346–53.

Adams, Kevin, and Marty Casstevens. 1996. *Wildflowers of the Southern Appalachians.* Winston-Salem, N.C.: John F. Blair.

Agrawal, Anurag, and Steven L. Stephenson. 1995. Recent successional changes in a former chestnut-dominated forest in southwestern Virginia. *Castanea* 60:107–13.

Alderman, J. Anthony. 1997. *Wildflowers of the Blue Ridge Parkway.* Chapel Hill: University of North Carolina Press.

Ambrose, J. P., and S. P. Bratton. 1990. Trends in landscape heterogeneity along the borders of the Great Smoky Mountains National Park. *Cons. Biol.* 4:135–43.

Amman, G. D., and C. F. Speers. 1965. Balsam wooly aphid in the southern Appalachians. *J. For.* 63:18–20.

Appalachian Mountain Club. 1977. *A.M.C. Field Guide to Mountain Flowers of New England.* Boston: Appalachian Mountain Club.

Bartram, William. [1791] 1955. *Travels of William Bartram.* Edited by Mark van Doren. New York: Dover Publications.

Baskin, Jerry M., and Carol C. Baskin. 1988. Endemism in rock outcrop plant communities of unglaciated eastern United States: An evaluation of the roles of the edaphic, genetic, and light factors. *J. Biogeogr.* 15:829–40.

———. 1992. Some considerations in evaluating and monitoring populations of rare plants in successional environments. *Natural Areas J.* 6:26–38.

Bell, C. Ritchie, and Anne H. Lindsey. 1990. *Fall Color and Woodland Harvests.* Chapel Hill, N.C.: Laurel Hill Press.

Benymus, Janine M. 1989. *The Field Guide to Wildlife Habitats of the Eastern United States.* New York: Simon and Schuster.

Berkley, Edmund, and Dorothy Smith Berkley. 1982. *The Life and Travels of John Bartram: From Lake Ontario to the River St. John.* Tallahassee: University Press of Florida.

Bieri, R., and S. F. Anliot. 1965. The structure and floristic composition of a virgin hemlock forest in West Virginia. *Castanea* 30:205–26.

Bolyard, Judith L. 1981. *Medical Plants and Home Remedies of Appalachia.* Springfield, Ill.: Charles C. Thomas.

Bowen, Brian. 1995. Purple loosestrife: An exotic invader of Tennessee's wetlands. *Tenn. Cons.* 61:28–30.

Bradley, Jeff. 1985. *A Traveler's Guide to the Smoky Mountains Region.* Boston: Harvard Common Press.

Bratton, S. P. 1974. The effect of European wild boar (*Sus scrofa*) on the high-elevation vernal flora in Great Smoky Mountains National Park. *Bull. Torr. Bot. Club* 101:198–206.

Braun, E. Lucy. 1950. *Deciduous Forests of Eastern North America.* New York: Hafner Publishing.

Brooks, Maurice. 1965. *The Appalachians.* Boston: Houghton Mifflin.

Brooks, R. R. 1987. *Serpentine and Its Vegetation: A Multidisciplinary Approach.* Portland, Ore.: Dioscorides Press.

Brown, Edward T., Jr., and Raymond Athey. 1992. *Vascular Plants of Kentucky.* Lexington: University Press of Kentucky.

Brown, Lauren. 1979. *Grasses: An Identification Guide.* Boston: Houghton Mifflin.

Canada Department of Forestry. 1961. *Native Trees of Canada.* 6th ed. Ottawa: Roger Duhamel.

Caplenor, Donald. 1965. The vegetation of the gorges of the Fall Creek Falls State Park. *J. Tenn. Acad. Sci.* 40:27–39.

———. 1979. Woody plants of the gorges of the Cumberland Plateau and adjacent Higland Rim. *J. Tenn. Acad. Sci.* 54:139–45.

Case, F. W., and R. B. Case. 1997. *Trilliums.* Portland, Ore.: Timber Press.

Catlin, David T. 1984. *A Naturalist's Blue Ridge Parkway Guide.* Knoxville: University of Tennessee Press.

Chapman, William K., and Alan E. Bassette. 1990. *Trees and Shrubs of the Adirondacks: A Field Guide.* Utica, N.Y.: North Country Books.

Chase, Jim. 1989. *Backpacker Magazine's Guide to the Appalachian Trail.* Harrisburg, Pa.: Stackpole Books.

Clebsch, E. E. C. 1989. Vegetation of the Appalachian Mountains east of the Great Valley. *J. Tenn. Acad. Sci.* 64:79–83.

Clebsch, E. E. C., and Richard T. Busing. 1989. Secondary succession, gap dynamics, and community structure in a southern Appalachian cove forest. *Ecol.* 70:728–35.

Coggins, Allen K. 1984. The early history of Tennessee's state parks, 1919–1956. *Tenn. Hist. Quart.* 43:295–315.

Constantz, George. 1994. *Hollows, Peepers, and Highlanders: An Appalachian Mountain Ecology.* Missoula, Mont.: Mountain Press.

Core, Earl L. 1966. *Vegetation of West Virginia.* Parson, W.Va.: McClain Printing.

———. 1981. *Spring Wildflowers of West Virginia.* Morgantown: West Virginia University Press.

Dann, Kevin T. 1988. *Traces on the Appalachians: A Natural History of Serpentine in Eastern North America.* New Brunswick, N.J.: Rutgers University Press.

Darlington, H. T. 1943. Vegetation and substrate of Cranberry Glades, West Virginia. *Bot. Gaz.* 104:371–93.

Duffy, David C., and Albert J. Meier. 1992. Do Appalachian herbaceous understories ever recover from clearcutting? *Cons. Biol.* 6:195–201.

Dugan, Patrick, ed. 1993. *Wetlands.* New York: Oxford University Press.

Duke, J. 1997. *The Green Pharmacy.* Emmaus, Pa.: Rodale Press.

Duncan, Wilbur H., and Marion B. Duncan. 1988. *Trees of the Southeastern United States*. Athens: University of Georgia Press.

Dwelley, Marilyn. 1976. *Spring Wildflowers of New England*. Camden, Maine: Down East Books.

———. 1977. *Summer and Fall Wildflowers of New England*. Camden, Maine: Down East Books.

Egler, Frank E. 1940. Berkshire Plateau vegetation, Massachusetts. *Ecol. Monog.* 10: 145–92.

Elias, Thomas S., and Peter A. Dykeman. 1982. *Field Guide to North American Edible Wild Plants*. New York: Outdoor Life Books.

Fernald, M. L. 1950. *Gray's Manual of Botany*. 8th ed. New York: American Book.

Foster, Steven, and Rogh Caras. 1994. *A Field Guide to Venomous Animals and Poisonous Plants*. Boston: Houghton Mifflin.

Foster, Steven, and James A. Duke. 1990. *A Field Guide to Medicinal Plants*. Boston: Houghton Mifflin.

Gaston, K. J., ed. 1996. *Biodiversity: A Biology of Numbers and Difference*. Oxford: Blackwell Scientific Publications.

Gattinger, Augustin. 1901. *Flora of Tennessee and Philosophy of Botany*. Nashville: Gospel Advocate Publishing.

Gleason, Henry A., and Arthur Cronquist. 1991. *Manual of Vascular Plants of Northeastern United States and Adjacent Canada*. 2d ed. Bronx: New York Botanical Garden.

Gore, Al. 1992. *Earth in the Balance: Ecology and the Human Spirit*. Boston: Houghton Mifflin.

Gupton, Oscar W., and Fred C. Swope. 1979. *Wildflowers of the Shenandoah Valley and Blue Ridge Mountains*. Charlottesville: University Press of Virginia.

———. 1981. *Trees and Shrubs of Virginia*. Charlottesville: University Press of Virginia.

———. 1987. *Fall Wildflowers of the Blue Ridge and Great Smoky Mountains*. Charlottesville: University Press of Virginia.

Hackney, D. T., S. M. Adams, and W. H. Martin, eds. 1992. *Biodiversity of the Southeastern United States: Aquatic Communities*. New York: Wiley.

Haragan, Patricia Dalton. 1991. *Weeds of Kentucky and Adjacent States: A Field Guide*. Lexington: University of Kentucky Press.

Hemmerly, Thomas E. 1989. Mistletoe parasitism in Tennessee. *J. Tenn. Acad. Sci.* 64: 121–22.

———. 1989. New commercial tree for Tennessee: Princess tree, *Paulownia tomentosa* Steud. (Scrophulariaceae). *J. Tenn. Acad. Sci.* 64:5–8.

———. 1990. *Wildflowers of the Central South*. Nashville: Vanderbilt University Press.

Hiers, J. K., and J. P. Evans. 1997. Effects of anthracnose on dogwood mortality and forest composition of the Cumberland Plateau (USA). *Cons. Biol.* 11:1430–35.

Hinkle, C. Ross. 1989. Forest communities of the Cumberland Plateau of Tennessee. *J. Tenn. Acad. Sci.* 64:123–29.

Hitchcock, A. S. 1971. *Manual of the Grasses of the United States*. 2d ed., revised by Agnes Chase. 2 vols. New York: Dover Publications.

303

Houk, Rose. 1993. *Great Smoky Mountains National Park.* Boston: Houghton Mifflin.

Hudson, Charles. 1976. *The Southeastern Indians.* Knoxville: University of Tennessee Press.

Hutchens, Alma R. 1973. *Indian Herbology of North America.* Boston: Shambala Publications.

Hutson, R. W., W. F. Hutson, and A. J. Sharp. 1995. *Great Smoky Mountains Wildflowers.* 5th ed. Northbrook, Ill.: Windy Pines Publishing.

Hyland, Fay, and B. Hoisington. 1981. *Woody Plants of Sphagnum Bogs of North New England and Adjacent Canada.* Bangor: University of Maine Press.

Johnson, Charles W. 1985. *Bogs of the Northeast.* Hanover, N.H.: University Press of New England.

Johnson, Edward A. 1992. *Fire and Vegetation Dynamics.* New York: Cambridge University Press.

Johnson, G. G., and S. Ware. 1982. Post-chestnut in the central Blue Ridge of Virginia. *Castanea* 47:329–43.

Jones, Ronald L. 1989. A floristic study of wetlands of the Cumberland Plateau of Tennessee. *J. Tenn. Acad. Sci.* 64:131–34.

Justice, William S., and Ritchie Bell. 1968. *Wildflowers of North Carolina.* Chapel Hill: University of North Carolina Press.

Keener, C. S. 1983. Distribution and biohistory of the endemic flora of the mid-Appalachian shale barrens. *Bot. Rev.* 49:65–115.

Keeney, Elizabeth B. 1992. *The Botanizers.* Chapel Hill: University of North Carolina Press.

Keever, Catherine. 1973. Distribution of major forest species in south-eastern Pennsylvania. *Ecol. Monog.* 43:313–27.

———. 1983. A retrospective view of old-field succession after 35 years. *Am. Midl. Nat.* 110:397–404.

Kephart, Horace. 1976. *Our Southern Highlanders.* Knoxville: University of Tennessee Press.

King, Duane H. 1979. *The Cherokee Nation: A Troubled History.* Knoxville: University of Tennessee Press.

Kricher, John C., and Gordon Morrison. 1988. *A Field Guide to Eastern Forests.* Boston: Houghton Mifflin.

Kuchler, A. W. 1964. *Potential Natural Vegetation of the Conterminous United States.* Amer. Geogr. Spec. Publ. No. 36.

Lindsay, Mary M., and Susan Power Bratton. 1979. Grassy balds of the Great Smoky Mountains: Their history and flora in relation to potential management. *Envir. Manag.* 3:417–30.

Lyon, John G. 1993. *Practical Handbook for Wetland Identification and Delineation.* Boca Raton, Fla.: Lewis Publishers.

MacDonald, Mart E. B. 1967. *Grassy Balds of the Great Smoky Mountains.* Knoxville: University of Tennessee Press.

Manning, Russ. 1981. The Big South Fork of the Cumberland: A new national river and recreation area. *Appalachia* (December):18–29.

———. 1993. *The Historic Cumberland Plateau: An Explorers Guide.* Knoxville: University of Tennessee Press.

Mansberg, L., and T. R. Wentworth. 1984. Vegetation and soils of a serpentine barren in western North Carolina. *Bull. Torr. Bot. Club* 111:273–86.

Mark, A. F. 1958. The ecology of the southern Appalachian grass balds. *Ecol. Monog.* 28:239–336.

Marsh, George P. 1965. *Man and Nature.* Cambridge: Harvard University Press.

Martin, W. H. 1992. Characteristics of old-growth mixed-mesophytic forests. *Nat. Areas J.* 12:129–35.

Martin, W. H., S. G. Boyce, and A. C. Echternacht, eds. 1993. *Biodiversity of the Southeastern United States: Lowland Terrestrial Communities.* New York: Wiley.

———. 1993. *Biodiversity of the Southeastern United States: Upland Terrestrial Communities.* New York: Wiley.

McCormick, J. Frank, and Robert B. Platt. 1980. Recovery of an Appalachian forest following the chestnut blight, or Catherine Keever—You were right! *Amer. Midl. Nat.* 104:264–73.

McGrath, Anne. 1981. *Wildflowers of the Adirondacks.* Utica, N.Y: North Country Books.

McIntosh, R. P., and R. T. Harley. 1964. The spruce-fir forests of the Catskill Mountains. *Ecol.* 45:314–26.

McKinney, Landon E. 1992. *A Taxonomic Revision of the Acaulescent Blue Violets (Viola) of North America.* Frankfort: Kentucky State Nature Preserves Commission.

Mooney, H. A., and J. A. Drake. 1986. *Ecology of Biological Invasions in North America and Hawaii.* New York: Springer-Verlag.

Mooney, James. 1982. *Myths of the Cherokee and Sacred Formulas of the Cherokee.* Reprint. Nashville: Elder Booksellers-Publishers.

Muller, Robert N. 1982. Vegetation pattern in the mixed mesophytic forest of eastern Kentucky. *Ecol.* 63:1901–17.

Oosting, H. J., and W. D. Billings. 1939. Edapho-vegetational relations in Ravenel's Woods, a virgin hemlock forest near Highlands, North Carolina. *Amer. Midl. Nat.* 22:333–50.

Peterson, Lee. 1978. *A Field Guide to Edible Wild Plants.* Boston: Houghton Mifflin.

Peterson, Roger Tory, and Margaret McKenney. 1968. *A Field Guide to Wildflowers.* Boston: Houghton Mifflin.

Petrides, George A. 1988. *A Field Guide to Eastern Trees.* Boston: Houghton Mifflin.

Pitillo, J. D., and T. E. Govus. 1978. *A Manual of Important Plant Habitats of the Blue Ridge Parkway.* Atlanta: U.S. Department of Interior, National Park Service, Southeast Regional Office.

Pritchard, Mack S. 1977. Exploring Savage Gulf—A last chance for wilderness. *Tenn. Cons.* 43:8–11.

Quarterman, Elsie, Barbara Holman Turner, and Thomas E. Hemmerly. 1972. Analysis of virgin mixed mesophytic forests in Savage Gulf, Tennessee. *Bull. Torr. Bot. Club* 99:228–32.

Radsford, Albert E., Harry E. Ahles, and C. Ritchie Bell. 1968. *Manual of the Vascular Flora of the Carolinas.* Chapel Hill: University of North Carolina Press.

Rheindardt, R. P., and S. A Ware. 1984. Vegetation of the Balsam Mountains of southwest Virginia: A phytosociological study. *Bull. Torr. Bot. Club* 111:287–300.

Rouleau, Raymond. 1990. *Petite Flore Forestière du Quebec.* Quebec: Ministère de l'énergie et des ressources.

Saunders, P. R., ed. 1980. *Status and Management of Southern Appalachian Mountain Balds.* Proceedings of a workshop, 5–7 November, Crossnore, N.C.

Shanks, R. E. 1954. Climates of the Great Smoky Mountains. *Ecol.* 35:354–61.

———. 1954. "Reference List of Native Plants of the Great Smoky Mountain." Manuscript.

Shelford, V. E. 1963. *The Ecology of North America.* Urbana: University of Illinois Press.

Shields, A. Randolph. 1981. *The Cades Cove Story.* Gatlinburg, Tenn.: Great Smoky Mountains Press.

Slack, Adrian. 1980. *Carnivorous Plants.* Cambridge: MIT Press.

Smallridge, Peter J., and Donald J. Leopold. 1994. Forest-community composition and juvenile red spruce *(Picea rubens):* Age-structure and growth patterns in an Adirondack watershed. *Bull. Torr. Bot. Soc.* 121:345–56.

Smith, Richard M. 1989. *Wild Plants of America.* New York: Wiley.

———. 1998. *Wildflowers of the Southern Mountains.* Knoxville: University of Tennessee Press.

Smith, William. 1981. *Air Pollution and Forests.* New York: Springer-Verlag.

Steele, Frederic L. 1982. *At Timberline: A Nature Guide to the Mountains of the Northeast.* Boston: Appalachian Mountain Club.

Stephenson, S. L. 1982. Exposure-induced differences in the vegetation, soils, and microclimate of north- and south-facing slopes in southwestern Virginia. *Va. J. Sci.* 33:36–50.

Strausbaugh, P. D., and Earl L. Core. 1954. *Flora of West Virginia.* Morgantown: West Virginia University Press.

Stupka, Arthur. 1964. *Trees, Shrubs, and Woody Vines of Great Smoky Mountains National Park.* Knoxville: University of Tennessee Press.

Swanson, R. E. 1994. *A Field Guide to the Trees and Shrubs of the Southern Appalachians.* Baltimore: Johns Hopkins University Press.

Threadgill, P. F., J. M. Baskin, and C. C. Baskin. 1981. The ecological life cycle of *Frasera caroliniensis,* a long-lived monocarpic perennial. *Amer. Midl. Nat.* 105:277–89.

Tilghman, N. G. 1989. Impacts of White-tailed Deer on forest regeneration in northwestern Pennsylvania. *J. Wildlife Management* 53:524–32.

U.S. Department of the Interior. 1991. *Proceedings of the Symposium on Exotic Pest Plants.* Washington, D.C.: U.S. Department of the Interior.

van Doren, Mark, ed. 1955. *Travels of William Bartram.* New York: Dover Publications.

Vogel, Virgil J. 1970. *American Indian Medicine.* Norman: University of Oklahoma Press.

Wegener, Alfred. 1966. *The Origin of Continents and Oceans.* New York: Dover Publications.

Weidensaul, Scott. 1994. *Mountains of the Heart: A Natural History of the Appalachians.* Golden, Colo.: Fulcrum Publishing.

Wells, B. W. 1937. Southern Appalachian grass balds. *Jour. Elisha Mitchell Sci. Soc.* 53:1–26.

Wells, James R. 1965. A taxonomic study of *Polymnia* (Compositae). *Brittonia* 17: 144–59.

Wharton, Mary E., and Roger W. Barbour. 1971. *A Guide to the Wildflowers and Ferns of Kentucky.* Lexington: University Press of Kentucky.

———. 1973. *A Guide to the Trees and Shrubs of Kentucky.* Lexington: University Press of Kentucky.

Whitney, G. G. 1990. The history and status of the hemlock-hardwood forests of the Allegheny Plateau. *J. Ecol.* 78:443–58.

Whittaker, R. H. 1956. Vegetation of the Great Smoky Mountains. *Ecol. Monog.* 26: 1–80.

Williams, John G., and Andrew E. Williams. 1983. *Field Guide to Orchids of North America.* New York: Universe Books.

Wilson, Edward O. 1992. *The Diversity of Life.* Cambridge: Harvard University Press.

Index